The Adaptiveness of IWRM: Analysing European IWRM research

The Adaptiveness of IWRM: Analysing European IWRM research

Edited by

Jos G. Timmerman, Claudia Pahl-Wostl, and Jörn Möltgen

Publishing

Published by IWA Publishing, Alliance House, 12 Caxton Street, London SW1H 0QS, UK

Telephone: +44 (0) 20 7654 5500; Fax: +44 (0) 20 7654 5555; Email: publications@iwap.co.uk
Web: www.iwapublishing.com

First published 2008

British Library Cataloguing in Publication Data
A CIP catalogue record for this book is available from the British Library

Library of Congress Cataloging- in-Publication Data
A catalog record for this book is available from the Library of Congress

ISBN: 1843391724

ISBN13: 9781843391722

Contents

Preface

PURPOSE OF THE BOOK

This book is developed by the European IWRM platform which was established under the EU FP6 research project NeWater (New Approaches to Adaptive Water Management under Uncertainty). It focuses on and discusses the results of research done on IWRM topics under EU research programmes but also considers developments in a wider context and the global scale, thus providing a state-of-the-art overview on the selected topics.

The book builds on the assumption that current IWRM in practice fails in promoting the implementation of cooperative and adaptive water management. This failure is attributed to the lack of a sound scientific base to understand the complexity of the water systems to be managed and to knowledge gaps in the understanding of the nature and the kind of processes of change needed to implement integrated water management approaches. The concept of adaptive management is considered an extension of IWRM which enables improved coping capabilities with the current issues water management is facing. Transition towards adaptive management investigates how different strategies, policies, and practices can be integrated to make management regimes more flexible and responsive to endogenously generated (e.g. population growth) and exogenously imposed (e.g. impacts of global climatic change) short- and long-term changes [1].

This book has two main goals:

(1) To learn about and capitalise on past and ongoing European research on IWRM.

(2) To learn and discuss the relationship between IWRM and Adaptive Water Management.

To accomplish the first goal, the book will provide an overview of the achievements drawn from research in Europe and globally of relevance for IWRM dealing with the topics (i) IWRM and Adaptive Water Management, (ii) participation in water management, (iii) transboundary regimes, (iv) economics of IWRM, (v) vulnerability and exposure to stress, (vi) understanding consequences of climate hazards and change, (vii) advanced monitoring systems, (viii) integration of IWRM and spatial planning in flood management, and (ix) social dimensions of IWRM. The book thus gives the reader a state of the art update on the scientific knowledge base supporting IWRM.

To accomplish the second goal, the book draws upon the IWRM achievements and gives an outlook on the readiness of IWRM to face the present-day challenges of uncertainty and complexity in water management. This provides insight into the necessary focus of EU-research in the near future and enables the development of policy recommendations. The recommendations are included in the first chapter.

[1] Pahl-Wostl, C. (2004) The implications of complexity for integrated resources management. *Environmental Modelling and Software*. **22**, 561-569.

SHORT INTRODUCTION TO THE NEWATER PROJECT

The project

The NeWater (New Approaches to Adaptive Water Management under Uncertainty) project is an Integrated Project under the 6th framework programme of the European Union. The project addresses some of the present and future challenges of water management. The NeWater project develops new methods for integrated water management taking into account the complexity of the river basins to be managed and the difficulty to predict the factors influencing them (e.g. climate, socio-economic developments). Water management must become more flexible and responsive to change to be able to cope with increasing uncertainty. Hence NeWater focuses in particular on the transition from current regimes of water management in a river basin to more integrated, adaptive approaches. The project recognises the value of highly integrated solutions and advocates integrated water resource management (IWRM) concepts. However, NeWater is based on the hypothesis that IWRM cannot be realised unless current water management regimes undergo a transition towards more adaptive water management.

The European IWRM platform

NeWater includes international and thematic platforms. The purpose of the European IWRM platform is to share experience and innovations in dialogues, publications and action, to further the European Research Area and to support the implementation of the Water Framework Directive and EU Water Initiative. One activity in the platform is to assess the current practice in IWRM and draw lessons for the transfer of new scientific methodologies for IWRM practitioners. The platform links the NeWater project to relevant past and ongoing EU projects, with the aim of providing an efficient mechanism through which the European and global IWRM community would be able to capitalise on existing results and achievements of these projects, minimise redundancies, and therefore further substantiate its added value to the European Research Area (ERA).

Preceding activities

In preparation of this book, several activities were completed. The first activity was a quick scan of EU research projects to identify the topics relevant for IWRM in the context of Adaptive Water Management (AWM). Following this, a survey among project coordinators was done to identify the maturity of the selected topics. Based on this survey and discussions during the 2005 Stockholm Water Week the following was concluded [2]:

(1) The level of maturity of issues in IWRM is varying;
(2) There is a considerable amount of EU funded research available on most of the topics;
(3) There is interest in synthesising the EU-funded research with respect to the topics. Such a synthesis should serve one or more of the following needs:
    (a) The increasing need from the research perspective to capitalise past EU-funded research.
    (b) The EC's need to feed discussions on research programmes (e.g. FP7)
    (c) The need of the application community to get access to research results.
    (d) The increasing need to support policy development by research findings.
    (e) The need to develop a comprehensive overview of the IWRM state of the art (based on EU funded projects).
(4) There are other relevant initiatives such as INCO-IWRM analysis and Harmoni-CA analysis;

The outcome of these activities was the selection of topics that are subject to the review represented in this book among EU research projects, a number of which were pre-selected, with the ultimate aim to develop a comprehensive overview of the state of the art on IWRM research which embodies the basis for further research.

INTENDED READERSHIP

The book has two primary target audiences. The first, and most important one, is the national and EU policymakers and practitioners who are or will become involved in the various aspects of IWRM, both on strategic and operational levels. The book will support them in gaining insight into the state of the art on IWRM as well as provide them with ideas on the required developments which will enable them to cope with the future water management challenges. The second

[2] Blind, M., Hattermann, F.F., Timmerman, J.G., and Sikorska, B. (2005) *European IWRM - research. Outline analysis of projects' contribution for the purpose of selecting topics for synthesis - A discussion starter to the European IWRM platform*. Report No 3 of the NeWater project.

target audience is the academic community concerned with the study of water management. The book provides them with a concise overview of available research and results. It also supports them in identifying future research topics.

ACKNOWLEDGEMENTS

The editors would like to thank the authors of the various chapters for their contributions to the authors meeting and their efforts in drafting the chapters, and also for reviewing chapters of other authors. Special thanks go to Bernd Hansjürgens (Helmholtz-Zentrum für Umweltforschung GmbH-UFZ), Fons Jaspers (Alterra, Wageningen, Netherlands), Edi Interwies (Interwies, Berlin, Germany), Nele Lienhoop (Helmholtz-Zentrum für Umweltforschung GmbH-UFZ), Willem Oosterberg (Institute for Inland Water Management and Waste Water Treatment, Lelystad, Netherlands) and Catharien Terwisscha van Scheltinga (Alterra, Wageningen, Netherlands) for their invaluable contributions to the discussions at the authors meeting and their review of the various chapters. The NeWater project is supported by the European Commission under the sixth framework programme (Contract no 511179 (GOCE)). The European Commission is gratefully acknowledged for their financial contribution.

Jos G. Timmerman, Claudia Pahl-Wostl, Jörn Möltgen

List of Contributors

Jeroen Aerts
Vrije Universiteit Amsterdam
Institute for Environmental Studies
De Boelelaan 1087
1081 HV Amsterdam, Netherlands

Rhodante Ahlers
UNESCO-IHE Institute for Water Education
Dept. of Management and Institutions
Westvest 7
P.O. Box 3015
2601 DA Delft, Netherlands

Natalia Alexeeva
Center for Transboundary Cooperation
ANO CTC-Spb
V.O. Kozhevennaya 34 –411
St-Petersburg 199106, Russia

Thomas Dworak
Ecologic - Institute for International and European Environmental Policy
Pfalzburger Strasse 43/44
D - 10717 Berlin, Germany

Raffaele Giordano
Water Research Institute – I.R.S.A.
Italian National Research Council – C.N.R.
Via De Blasio 5
70123 Bari, Italy

Geoffrey D. Gooch
Professor of Political Science
Jean Monnet Chair of European Political Integration
Department of Management and Engineering
Linköpings University
58183 Linköping, Sweden

Ing-Marie Gren
Department of Economics
SLU
P.O. Box 7013
750 07 Uppsala, Sweden

Bernd Hansjürgens
Helmholtz-Zentrum für Umweltforschung GmbH-UFZ
Permoserstraße 15
04318 Leipzig, Germany

Fred F. Hattermann
Potsdam Institute for Climate Impact Research
P.O. Box 60 12 03
14412 Potsdam, Germany

Dave Huitema
Vrije Universiteit Amsterdam
Institute for Environmental Studies
De Boelelaan 1087
1081 HV Amsterdam, Netherlands

Eduard Interwies
InterSus - Sustainability Services
Chodowieckistr. 2
10405 Berlin, Germany

Fons Jaspers
Alterra
P.O. Box 47
6700 AA Wageningen, The Netherlands

Paul Jeffrey
Centre for Water Science
Cranfield University
Cranfield, Beds MK43 0AL UK

Susanne Kadner
Potsdam Institute for Climate Impact Research
P.O. Box 60 12 03
14412 Potsdam, Germany

Valentina Krysanova
Potsdam Institute for Climate Impact Research
P.O. Box 60 12 03
14412 Potsdam, Germany

Anna Leipprand
Ecologic - Institute for International and European Environmental Policy
Pfalzburger Strasse 43/44
D - 10717 Berlin, Germany

Nele Lienhoop
Helmholtz-Zentrum für Umweltforschung GmbH-UFZ
Permoserstraße 15
04318 Leipzig, Germany

Hans de Moel
Vrije Universiteit Amsterdam
Institute for Environmental Studies
De Boelelaan 1087
1081 HV Amsterdam, Netherlands

Jörn Möltgen
Institute for Environmental Systems Research
University of Osnabrück
Barbarastr. 12
49076 Osnabrück, Germany

Willem Oosterberg
RWS Centre for Water Management
P.O Box 17
8200 AA Lelystad, Netherlands

Claudia Pahl-Wostl
Institute for Environmental Systems Research
University of Osnabrück
Barbarastr. 12
49076 Osnabrück, Germany

Joachim Post
German Aerospace Center
51170 Köln, Germany

Dagmar Ridder
Institute for Environmental Systems Research
University of Osnabrück
Barbarastr. 12
49076 Osnabrück, Germany

Gulnara Roll
Political Science Department
University of Tartu
Tiigi 78 - 203
50410 Tartu, Estonia

David Sauri
Department of Geography and Institute of Environmental Science and Technology
Universitat Autònoma de Barcelona
08193 Bellaterra, Spain

Catharien Terwisscha van Scheltinga
Alterra
P.O. Box 47
6700 AA Wageningen, Netherlands

Jos G. Timmerman
RWS Centre for Water Management
P.O Box 17
8200 AA Lelystad, Netherlands

Vito F. Uricchio
Water Research Institute – I.R.S.A.
Italian National Research Council – C.N.R.
Via De Blasio 5
Bari, Italy

Michele Vurro
Water Research Institute – I.R.S.A.
Italian National Research Council – C.N.R.
Via De Blasio 5
Bari, Italy

Markus Wrobel
Potsdam Institute for Climate Impact Research
P.O. Box 60 12 03
14412 Potsdam, Germany

1

The adaptiveness of IWRM

Jos G.Timmerman, Claudia Pahl-Wostl and Jörn Möltgen

1.1 IWRM IN A CHANGING WORLD

Water is an essential natural resource. There is plenty of freshwater in the world, but it is not distributed evenly in space and in time. Many people have too little water to grow their crops, while on the other hand floods frequently threaten lives and harvests. Also, the quality of the available water is often poor and access to safe drinking water is lacking (Cosgrove and Rijsberman 2000; Hinrichsen 2003; Saeijs and van Berkel 1995; Serageldin 1994).

Current water resource management poses many challenges. Water managers need to solve a range of interrelated water dilemmas, such as balancing water quantity and quality, flooding, drought, maintaining biodiversity and ecological functions and services, in a context where human beliefs, actions, and values play a central role. Furthermore, the growing uncertainties of global climate change and the long-term implications of operational management actions are additional hurdles for a sustainable water management in view of an increasing water scarcity. Policy development must therefore aim at developing in such a way that the present needs are met without compromising the ability of the future generations to meet their needs (Rauch 1998). Sustainability in this sense requires balancing of needs including future needs which are not really predictable.

Water problems and water management options are as much a product of the social, economic and institutional context as they are of the biophysical and technical factors governing local hydrological conditions (Gleick 2003). Many water related challenges have to do with socio-economic distribution and access, especially in developing regions. Natural changes must in this context be distinguished from human-induced changes in the water system (Falkenmark 2000). Since water is the cornerstone of most economic activity, equitable distribution of supplies under changing conditions is often more of a challenge than absolute limitations on the available resource (Pahl-Wostl *et al.* 2005).

Water management is currently seen in terms of 'emergent complexity' (Ingram 2001). Next to the 'tame' problems there is now a class of problems termed 'wicked' or 'persistent'. Tame problems are not trivial problems, but problems which can be tackled with some degree of confidence. Wicked problems on the other hand, are unstructured problems which are characterised by a high level of technical and strategic uncertainties (there is no objective measure of success), and which have to be solved in a situation in which the parties involved are highly dependent on each other while each party has its own moral, political or professional preferences (Buckingham Shum 1997; Klijn *et al.* 2000; Pahl-Wostl 2007a; van der Brugge *et al.* 2004).

Integrated Water Resources Management (IWRM) is for a long time considered to be able to cope with complex water management problems. IWRM accounts for social, economic and environmental factors and integrates surface water, groundwater and the ecosystems through which they flow (GWP-TAC 2000). Current water management, based on the concept of IWRM, however fails to satisfactorily solve the problems it is facing (for instance Falkenmark 2000; IWA 2002; Kabat and van Schaik 2003). It is increasingly argued that IWRM should be based on an Adaptive Management (AM) approach, which is defined as a systematic process for improving management policies and

© 2008 IWA Publishing. *The adaptiveness of IWRM; Analysing European IWRM research* Edited by Jos G. Timmerman, Claudia Pahl-Wostl and Jörn Möltgen. ISBN: 9781843391722. Published by IWA Publishing, London, UK.

practices by learning from the outcomes of implemented management strategies (for instance Medema *et al.* 2007; Pahl-Wostl 2007a). This book investigates the state of the art in IWRM on selected topics to conclude on the necessary conditions to cope with the complexity in water management.

This chapter introduces the main concepts which are dealt with within this book, namely the concept of transition to adaptive water management, water management regimes, governance, vulnerability, the social dimension of water management, and participation. A more in-depth discussion of the concepts can be found in the respective chapters. This chapter will next to that discuss the main outcomes from the review of IWRM research as discussed in this book and the conclusions and recommendations which are be drawn from the book.

1.2 CONCEPTS IN WATER MANAGEMENT

1.2.1 The transition towards adaptive water management

Major challenges faced in water management are inherent in the variability and changing nature of water supplies as a result of climate change, along with the limited nature of scientific information and technical knowledge. Reliance on conventional methods of water management, based on the statistical analysis of historical data series, is in many cases not sufficient. Under such conditions, analysis must proceed iteratively with an emphasis on uncertainties rather than on the known (Kabat and van Schaik 2003; Pahl-Wostl 2002). Adaptive management is advocated as a timely extension of IWRM to cope with these challenges. Adaptive management aims at increasing the adaptive capacity of river basins based on a profound understanding of key factors that determine a basin's vulnerability. More attention has to be devoted to understanding and managing the transition[1] towards more adaptive regimes that take into account environmental, technological, economic, institutional and cultural characteristics of river basins (Pahl-Wostl *et al.* 2005; Pahl-Wostl 2007b). This transition implies a change towards understanding management as learning rather than control, and towards including the human dimension as integral part of the management process (Gleick 2003).

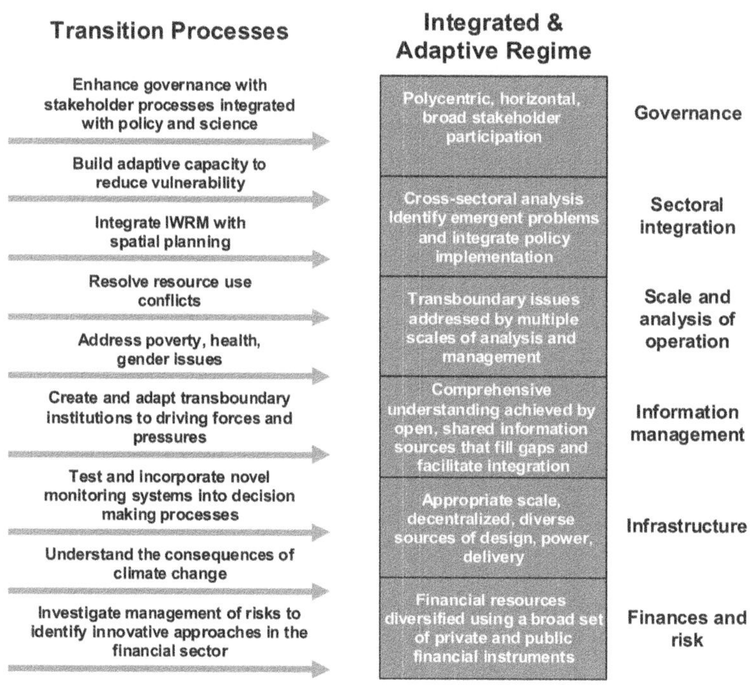

Figure 1: The transition towards adaptive regimes

Figure 1 represents essential elements of the transition. It identifies the key areas of policy oriented research required to understand the transition process towards an adaptive regime. The integration of these key areas will be essential for success (Pahl-Wostl *et al.* 2005). The figure represents elements of adaptive regimes and what is needed for the transition into integrated and adaptive water management as will be discussed in Chapter 2. This includes enhancing governance with stakeholder processes which integrate between policy and science involving better understanding on public participation in general and inclusion of informal institutions. Chapter 3 deals with this. As water management

[1] A transition is a structural change in the way a societal system operates (van der Brugge *et al* 2004).

should be focused on river basins, it must adapt its transboundary institutions or create them in order to be able to adjust to the relevant drivers and pressures (Chapter 4). The financial and legal arrangements must be such that they include and enhance innovative approaches to risk management (Chapter 5). Also, the management system must build in adaptiveness to reduce vulnerability and exposure to stress (Chapter 6). This can, among others, be achieved by closely linking spatial planning to IWRM (Chapter 9). For all this it is necessary to better understand the consequences of climate hazards and change (Chapter 7). Furthermore, advanced monitoring systems are needed which are better integrated in the decision-making processes (Chapter 8). Finally, in the transition to adaptive water management, ample attention is needed for poverty, gender and health issues affecting the social dimension of sustainability (Chapter 10).

The central tenet of our argumentation concerns the transition towards more integrated and adaptive regimes of river basin water management in the future. This transition, in general, calls for a highly integrated water resources management concept. The need for greater integration is now widely recognised and accepted but there is, in actuality, little theoretical foundation as well as practical experience regarding how this integration could be achieved or of the consequences of attempts to achieve it.

1.2.2 Water management regimes

Regimes are defined in many different ways. Krasner (1983), for instance, defines a political regime as a set of institutions: 'implicit or explicit principles, norms, rules, and decision making procedures around which actors' expectations converge in a given area of international relations'. Pahl-Wostl defines it wider as the whole complex of technologies, institutions, environmental factors and paradigms which are highly interconnected and together form the base for the functioning of the management system targeted to fulfil a societal function (Pahl-Wostl 2007b). In this chapter, we will start off from the wider perspective. In Chapter 4, a more in depth discussion will be presented.

Regimes play a role at different levels. For water management the (transboundary) river basin level seems to be appropriate. A river basin management regime comprises the whole of actor networks, water law, water policy, information management and finances, because these elements reflect the principles, norms, rules and decision-making procedures in a concrete, analysable way (Raadgever *et al.* 2007). These elements will shortly be discussed here.

Water management has to build on both formal and informal networks. Current (transboundary) water management often centres around national governments and can be described as a 'prediction and control' regime (Moberg and Galaz 2005; Pahl-Wostl 2002). This system's approach is derived from a mechanistic thinking in which system behaviour and response can be predicted and optimal control strategies can be designed. However, it is questionable whether such an approach can provide the base for an effective integrated and adaptive water management coping with complexity and change. To better cope with the water management challenges, cooperation should be established between different levels of river basin administration. Furthermore, broad interaction and cooperation with NGO's, experts (e.g. scientists), and citizens is needed to exchange (new) knowledge and cooperation in framing, reframing, and solving problems (Ridder *et al.* 2005) so that all stakeholders continuously and actively learn (Folke *et al.* 2005; Pahl-Wostl and Hare 2004). Such interaction can be formalised to some extent in law or organisational structures or can take place on a purely informal basis (for instance Huitema *et al.* 2007). The interactions make it possible that conflicts are dealt with constructively, resulting in inclusive agreements to which the parties are committed (Raadgever *et al.* 2007).

The legal framework can support water management in different ways. Water law should include arrangements about integrated water management, governance networks, public participation, information management, planning and implementation, and financial aspects. To become adaptive, legislation should provide incentives to adapt the management regime to changing circumstances as well. It should therefore include provisions to review and change policies and regulations periodically (Raadgever *et al.* 2007).

Water policies are the formal documents which prescribe current and future water management (Raadgever *et al.* 2007). Water policies have until recently relied on technological possibilities, often large-scale infrastructure (dams, reservoirs, centralised waste water treatment plants) supposed to guarantee long-term stability and reliability and a higher efficiency in implementation, maintenance and operation than de-centralised small-scale structures. Due to long-life times and large sunk costs, change is largely impossible once a certain infrastructure is in place (Pahl-Wostl 2007b). Also in current water management regimes responsibilities are often fragmented with little interaction between areas such as flood protection, regional planning, waste water treatment, water supply, or hydropower generation. This allows for a high specialisation of dealing with a single problem but prevents integration and also prevents change and learning (Pahl-Wostl 2007b). From an adaptive management point of view, the full range of possible measures should be considered, in the context of several scenarios for future developments (Raadgever *et al.* 2007).

Information management deals with the iterative process of determining information needs and collecting, interpreting, exchanging and using information. In current water management regimes, technical experts define information needs. The main data collection and regular monitoring focuses on the state of the environment. Knowledge is not shared and communication with stakeholder and interest groups is mainly by passive channels (cf. reviews and case studies in Timmerman and Langaas 2004b). The criteria for adaptive information management are strongly connected to those for the actor networks and for policy development. A broad range of governmental and non-governmental stakeholders should have the possibility to express their information needs, direct the information production, and exchange and discuss data and viewpoints. Also, information should not only reflect multiple perspectives, but should also consider current and future uncertainties (Giordano *et al.* 2007; Timmerman and Langaas 2005). In order to minimise the risk of misinterpretation or misuse of information, clear communication about the interpretations and assumptions which have been made is essential (Gooch 2004). This requires critical (self-)reflection of the individual frames of interpretation (Ridder *et al.* 2005).

In a transboundary context, it is important that national governments and transboundary commissions exchange information and actively disseminate information to the public (Nilsson 2003). In addition, effective information management at the local level is crucial in order to facilitate flexible management approaches. Strategic information management, e.g. using information as for instance a means to postpone or to direct decision-making, in transboundary cooperation should be limited as this reduces trust and cooperation. Information production and exchange should result in the use of information in policy debates, and should influence river basin management decisions (Timmerman and Langaas 2004b).

An appropriate financing system requires sufficient resources to be available for water management, from private as well as public sources. The supposition should be that the additional investment on the short term results in larger costs savings on the longer term. Aspects of financing include cost recovery from the users of the river who are considered to benefit from adaptive management, because it makes them aware of the consequences of their activities and helps to avoid overexploitation of the river basin. Also, to ensure actual implementation of measures, decision-making and financing should be in the hands of the same party (Raadgever *et al.* 2007).

1.2.3 Governance

The introduction of the term governance over the last few decades has signalled a change in thinking about the nature of policies. The notion of government as the single decision making authority has been replaced by multi-scale, polycentric governance taking thus into account that a large number of stakeholders in different institutional settings contribute to policy and management of a resource. This change reflects a more general trend in public policy, different from the old hierarchical model in which state authorities exert sovereign control over the people and groups making up civil society (Mayntz 1998). 'Governance' takes into account the increasing importance of basically non-hierarchical modes of governing, where non-state, private corporate actors (formal organisations) participate in the formulation and implementation of public policy (Rhodes 1997). Governance encompasses thus a broad range of processes related to the coordination and steering of a wide range of actors by formal and informal institutions. Governance theory includes the study of principles, norms and rules, of informal institutions, and flexible patterns, which are constantly forming and reforming (Smouts 1998), such as those in water policy.

A governance system is based on voluntary agreements which are confined through a set of rules that are binding in the sense that they create convergent expectations and govern behaviour. This contrasts with a government system, where obligations emanate from hierarchical norm- and rule-setting processes (Mayer *et al.* 1993) (also see Chapter 3). Effective modern governance can therefore only emerge in societies which meet the following institutional and structural preconditions, both on the side of the political regime and on the side of civil society: (1) political authorities must be powerful, but not omnipotent; (2) the existence of a strong, functionally differentiated, and well organised civil society; (3) there must also be corporate actors who represent different functional as well as different socio-economic interests; and (4) there must exist among the different social groups and organisations at least a minimal sense of identification with, and responsibility for, the greater whole, in short, a common identity (Mayntz 2003). Moving towards more adaptive water management demands moving towards an improved system of water governance, which entails building on these aspects (see Chapters 2 and 3).

1.2.4 Vulnerability

Vulnerability is defined in many different ways in literature (Adger 2006). Definitions of vulnerability in the climate change related literature tend to fall into two categories, viewing vulnerability either (i) in terms of the amount of (potential) damage caused to a system by a particular climate-related event or hazard, or (ii) as a state which exists within a system before it encounters a hazard event. The former view has arisen from an approach based on assessments of hazards and their impacts, in which the role of human systems in mediating the outcomes of hazard events is downplayed or neglected. This vulnerability as a function of hazard, exposure and sensitivity, is referred to as biophysical vulnerability. Biophysical vulnerability is concerned with the ultimate impacts of a hazard event, and is often viewed in terms of the amount of damage experienced by a system as a result of an encounter with a hazard (Brooks 2003).

Conversely, the view of vulnerability as a state (i.e. as a variable describing the internal state of a system) has arisen from studies of the structural factors which make human societies and communities susceptible to damage from external hazards. In this formulation, vulnerability is something which exists within systems independently of external hazards. For many human systems, vulnerability viewed as an inherent property of a system arising from its internal characteristics may be termed 'social vulnerability' (Adger 1999). Social vulnerability is determined by factors such as poverty and inequality, marginalisation, food entitlements, access to insurance, and housing quality. It is social vulnerability that has been the primary focus of field research and vulnerability mapping projects, which are generally concerned with identifying the most vulnerable members of society, and examining variations in vulnerability between or within geographical units which may experience similar hazards. In this formulation, it is the interaction of hazard with social vulnerability that produces an outcome, generally measured in terms of physical or economic damage or human mortality and morbidity (Downing *et al.* 2006). Hence social vulnerability may be viewed as one of the determinants of biophysical vulnerability (Brooks 2003). Chapter 6 provides an in-depth discussion on vulnerability.

1.2.5 Social dimensions (poverty / gender)

Poverty is a complex phenomenon with different characteristics in different places. Three key dimensions to poverty reduction can be distinguished: livelihoods, health and vulnerability (World Bank 2002). In any competition over access to resources, whether these be natural resources or man made services and livelihood opportunities, those living in poverty do less well than others. 'Water poverty' is an important and unrecognised component of poverty generally, and lack of access to water resources and adequate services is integral to the disadvantaged situation of the vast majority of the poor (GWP-TEC 2003). The lack of access to clean water and proper sanitation is among the prime causes of people being annually affected by water-borne diseases, primarily in the developing countries. The poorest segments of society and children are hit hardest (Jønch-Clausen 2004). Poverty is also an important aspect of vulnerability because of its direct association with access to resources, which affects both baseline vulnerability and coping from the impacts of extreme events. Poorer people tend to live in more 'marginal' and more hazardous areas, though the causality in this relationship is difficult to determine. Location affects the elements of poverty: in economic terms marginal areas have higher marginal costs of access. Transport to centres of distribution of government social security at times of hazard impacts, and the higher exposure of marginal areas to hazards, such as poor housing susceptible to earthquake damage or land prone to flooding, are both elements in this spatial vulnerability-poverty interaction (Adger 1999).

Water management should facilitate management of water resources and water services in ways which will help to reduce poverty (GWP-TEC 2003). IWRM however tends to focus on issues such as cost recovery, reallocation of water to 'higher value' uses, and environmental conservation (Merrey *et al.* 2005). Adaptive water management should therefore focus on the equitable allocation of benefits from water and services dependent on it, for healthier and more productive lives among the most at-risk and disadvantaged population groups. Chapter 10 will describe in detail how this can be achieved.

1.2.6 Participation

Water management is faced by problems arising from the production of scientific information, the communication of that information, legal systems, and institutions (Gooch and Stålnacke 2006). This has lead to an increased awareness of the necessity of including stakeholder and public participation in water management decision-making, as well as in other areas (OECD 2001). The motivation for this involvement is that it is expected to lead to greater legitimacy for the policies (due to the spread of responsibility), and to improvements in policy implementation (due to a readier acceptance of policies in which stakeholders and the public have participated) (Lise *et al.* 2004). Yet another reason for examining ways to increase participation is the fact that large sectors of the public 'have grown distrustful of politicians, sceptical about democratic institutions, and disillusioned about how the democratic process functions' (Dalton 2004). Current trends indicate that three processes are ongoing, which have caused the growing importance of public participation (Lise *et al.* 2004):

(1) *Negativism.* There is a general feeling of democratic deficit. On the one hand, some governments seem no longer to deliver the results as desired by the public and public trust in political decisions has fallen dramatically and is not recovering, causing a crisis of legitimacy. On the other hand, governments are constrained by global economic interests and by multi-lateral obligations, and lack the capacity to meet the many and frequently conflicting local population needs.

(2) *Pragmatism/efficacy*. There is a growing awareness among governments that decisions are often no longer acceptable without participation of the public in the decision-making process. Without this public consensus, decisions may fail.

(3) *Citizenship/sustainability*. Nowadays people want to be able to shape their own futures. We have a self-evolving society, which is a recent trend. Information and Communication Technology (ICT) has a role in this, in that it opens up information to an ever-wider audience and enables people to coordinate and direct actions. It is now possible to visualise images of future flooding, or landscapes or coastal patterns to allow stakeholders to see for themselves how future patterns of landscape and policy may evolve. Such images are critical in the participatory process.

The usefulness of contributions for water managers may vary but the increased levels of education are undeniable. And even without the benefits of formal education, citizens often posses 'local knowledge' or 'ordinary knowledge' which relates to their being on their land and which water managers need to make accurate decisions. Faced with an increasingly complicated and uncertain environment, there is an increasing realisation of the need to learn to manage together (Ridder *et al.* 2005) and participation is consequently intertwined with the concept of Governance (Chapter 3).

1.3 CONCLUSIONS

This section provides an overview of the main findings from the chapters in this book. The section is arranged according to the framework introduced in Section 1.2.1 to put the separate conclusions of the chapters in a wider context. This will lead to an overview of further research needs in the following section. More specific conclusions and recommendations are included in the respective chapters.

1.3.1 Governance

From Chapters 2, 3 and 4 it becomes clear that the water management regime is a pivoting point in achieving a well-organised water governance system which supports adaptive water management. New management paradigms are needed to develop such new water governance. Space is needed for instance for creative and out-of-the-box thinking, and the mechanistic worldview is to be replaced by a more organic, system-based perspective which recognises dependencies and evolutionary dynamics (Chapter 2). Furthermore, water managers and authorities need to assess and specify their aims and ambitions (Chapters 3 and 10) while mechanisms must be created to include the various views into the decision-making and policy process (Chapter 3). Also, civil society, or its representatives, must have the necessary information and knowledge to participate, entailing an information management system which involves all stakeholders at their disposal (Chapter 8).

This entails that authorities must develop mechanisms which incorporate participation both in terms of input (who participates when) as well as output (what is done with the results) to facilitate this water governance. Civil society in such a regime must be enabled to raise its concerns, for instance on the issue of vulnerability (Chapter 6 and 7) or social inequity (Chapter 10). Linked to that is the possibility for the public at large to gain insights from available information (Chapter 8). All this requires an increase in flexibility and adaptive capacity of the water management regime and building and sustaining of trust and social capital to support this collaborative governance system.

However, especially in transboundary regimes where disputes over water exist and cooperation between riparian countries is not established or where cooperation is built on power or interests, such a flexible and adaptive capacity is not easily established, as discussed in Chapter 4. Disputes over water most likely appear in basins whose management institution was developed under a single jurisdiction, but which was altered or shattered such that jurisdiction suddenly became divided among two or more nations, and in the case of unilateral development in the absence of a treaty or commission (Wolf *et al.* 2003). Examples of such situations are included in the case studies of the NeWater project and some of these regimes are considered less adaptive (Raadgever *et al.* 2007). As Chapter 4 discusses, the concept of adaptive management is built on the school of the cognitive perspective towards international regimes, where a neoliberal perspective in regions where cooperation is not well established provides a better assessment of the situation and as a consequence would better support research and knowledge building on adaptive management in such a situation. Research on adaptive water management should therefore concentrate more on the capacities of countries to build and sustain transboundary regimes, with special attention for basins over which disputes exist.

1.3.2 Sectoral integration

IWRM is not a technological management issue but must take place within its wider social and economic context. Chapter 7 for instance clearly shows the linkages between water management, agriculture, demographic developments, and the water sector while Chapter 9 shows how spatial planning should go hand in hand with IWRM. Referring specifically to impacts of climate change on water management, from Chapters 6 and 7 we find that these impacts are not only connected to physical conditions and technological solutions to minimise exposure relative to the magnitude of natural hazards, but that they are as well linked to the resistance and resilience of societies that are struck by a natural hazard. In this context, Chapter 6 portrays the social and economic aspects of the concept of social vulnerability, while Chapter 10 describes water security in terms of income and gender. There is consequently a wide range of drivers which influence IWRM and the political process which shapes it.

The consequences of this finding are manifold, as new ways have to be found to deal with these different and contradictory needs. These contain aspects such as connecting concepts from different disciplines, and enabling decision makers as well as civil society to cope and balance with these concepts and the knowledge associated with them. Information management as input to the decision-making process has to be able to connect the different types of information and also draw from local knowledge (Chapter 8), while models used in the different disciplines must be able to connect to each other and provide integrated outputs (Chapter 2).

To ensure a better integration between the different disciplines and water sectors, detailed empirical research is needed on the interlinkages between the sectors and disciplines. As for instance described in Chapter 5, perceptions of risk and uncertainty can lead to unexpected outcomes in the economic discipline. Reviewing issues should therefore be done on the basis of ecological, social and economic perspectives. The approach towards vulnerability as *social vulnerability* (Downing *et al.* 2006) is a good example of such an approach, which however still needs to be put into the day-to-day water management practice.

1.3.3 Scale and analysis of operation

Even within the context of integrated water management, many water managers remain focused on opportunities for technical solutions and supply development, linked to a centralised view on governance. Water development is in this view generally a large-scale construction activity requiring major resources and organisation, which are both often the prerogatives of large, state-run types of governance structures. These characteristics have contributed to the development of much large, centralised water infrastructure around the world. Efficient water use and equitable allocation however largely depend on the behaviour of the individual users (ranging from the individual farmer or domestic consumer to users of very large quantities, such as major hydropower facilities), which is an important driver for the development of multi-scale, polycentric governance as discussed in Section 1.2.3. Efficient and equitable management in a situation of increasing uncertainty, will have to be an inherently local activity where courses of action and the incentives to undertake them are contingent on specific local hydrological, economic, technical and social conditions (Pahl-Wostl *et al.* 2005).

The concept of social vulnerability markedly shows how the management system operates at a range of different scales, both in time as well as spatially. The natural hazard can occur at a small spatial and time scale in the form of flash floods or at large time and spatial scale like in a change of average seasonal temperature or precipitation. The social and economic responses to these events also work on different scales like local and national scale. For instance market intervention at the national or international level may affect the price of a commodity produced by a household or community, with dramatic consequences for the latter's economic status and resulting resistance and resilience. Adaptation to environmental risks at the local scale often reduces the vulnerability of those best able to mobilise resources, rather than the most vulnerable (Adger 2006). Systems can therefore not be treated as closed, nor can a system's ability to adapt be assessed without considering the role of obstacles to adaptation that might be determined by processes operating outside of the system in question (Brooks 2003). Chapters 6, 7 and 9 go into more detail on these issues.

To address the different aspects at the appropriate scales requires analysis possibilities on these very scales. Local and regional water resources management should therefore be combined with integrative river basin approaches and has to be embedded into a perspective of global change (Pahl-Wostl *et al.* 2005). Evidence is accumulating that more consciously addressing scale issues and the dynamic linkages across levels are more successful at assessing problems and finding solutions which are more politically and ecologically sustainable (Cash *et al.* 2006). The question we consequently have to deal with in view of scaling and analysis of operations is hence how to integrate management approaches at local, regional and basin wide scales? Scaling issues need to be explored to understand the complex dynamics of institutional resource regimes and to improve the match between biophysical and actor based scales. The latter also relates to developing regimes which , for instance, deal with implementation of EU policies versus regimes dealing with water management of a specific river basin (see Chapter 4).

1.3.4 Information management

As described in the previous sections, the complexity of water management and the increase in uncertainties in the light of global change, along with globalisation, which is increasingly influencing river-basin level water management, requires new approaches in governing water management and integrating over different sectors and scales. For example, the need to make large investments to better cope with the impacts of climatic extremes cannot as yet be substantiated in a fully deterministic way as this would require the provision of exact information on the frequency and magnitude of climatic extremes in the future (Pahl-Wostl *et al.* 2005). All this increasing need for knowledge of the complexity of the processes in water systems has led to a growing demand for environmental information (Timmerman *et al.* 2000).

Sharing of data and information is a key issue, even more in transboundary regimes (Timmerman and Langaas 2004a), while in many cases the information which would be required for (adaptive) water management is not available. This, among others, requires the development of a new generation of monitoring systems and close involvement of stakeholders in data collection and monitoring (Pahl-Wostl *et al.* 2005). While several scholars (among others Allen *et al.* 2001; Grossmann 2005; Mostert 2005; Savenije and van der Zaag 2000) point out that information has major impact on policy- and decision making and consequently plays a much more prominent role than currently recognised, in European research information management is still a subject that is little studied and generally considered as a technical solution (see Chapter 8). First steps are underway as for instance a novel, adaptive monitoring system is under development within the NeWater project (Giordano *et al.* 2007), but concrete results from this study are not available yet.

Summarising, information issues which water management is facing are (Chapter 4):

- Changes in governance that requires a wider audience to be provided with solid information regarding water management;
- The need to combine various types of information stemming from various sectors and disciplines, and covering different spatial and temporal scales;
- Novel indicators that must be developed and tested to be able to measure and monitor distinct water management aspects like for instance valuation of large-scale aquatic systems, societal costs of measures, or resilience of a civic community;
- Uncertainties in water management, especially regarding climate change, which need to be reduced or at least quantified;
- Integration of scientific knowledge with local knowledge to account for local peculiarities.

1.3.5 Infrastructure

As already discussed in Section 1.3.3, in the current management regime, technical infrastructure is designed to be rigid and centrally controlled. These characteristics have contributed to the development of much larger, centralised water infrastructures around the world. This applies for example to the design of dikes for flood management or of water supply systems to meet peak demand at any time. One consequence of implementing big infrastructures is that huge sunk costs make these systems extremely inflexible. Such infrastructures are becoming inherently unsuited to support many water management functions. Different designs of infrastructure would be required under an adaptive management regime where technical infrastructure combines centralised and decentralised systems in a multi-scale, modular approach. This gives the systems the characteristics of a complex adaptive system, with internal degrees of freedom and distributed control which have a higher capacity to buffer variability in both the socioeconomic and natural environment (Pahl-Wostl *et al.* 2005).

The approach is highlighted in Chapters 6 and 9 where the risks of natural hazards are portrayed on different scales which do not always match the infrastructure in place. Moreover, the most vulnerable areas are not always the areas which are best protected. To enable a better-differentiated, decentralised infrastructure, improved assessment of vulnerability and risks is needed. To this purpose, models that predict the effects of climate change must support integrated assessments on a smaller scale than currently available (Chapter 2). Research should support development of such fine-tuned assessments.

1.3.6 Finances and risk

Chapter 5 shows how economic approaches can support management decisions and risks related to them. Economists generally approach the problem of decision-making under uncertainty via the expected utility approach. This approach supposes that a risk is evaluated in terms of the probability of certain outcomes. The approach therefore requires specification of the probabilities attached to different outcomes. If these probabilities are not known, it is assumed that people form subjective judgements about them, and then maximise the expected utility as if they were known (Perrings 2005).

Developments in especially cost-benefit analyses have, despite these limitations, provided a solid base to support policy decisions on a national scale. Issues like equity and fairness, flexibility, or reduction of transaction costs, which are essential elements for adaptive approaches, however gained less attention. Also, closed-economy models have dominated environmental economic analysis for a long time and as a consequence, international aspects of environmental problems are often not covered at all, or dealt with only as an afterthought and in a rather perfunctory manner (Schulze and Ursprung 2001). These issues, especially treated in a transboundary context need more focus in European research.

Next to that, a serious flaw of the expected utility approach is that low probability events will attract little attention, even if the value of the outcome is large. There is considerable evidence that decision-makers do not treat low probability-high cost outcomes as the decision model predicts and that decision-makers generally underestimate risks from frequent causes and overestimate risks from infrequent causes (Perrings 2005). Perceptions of risk and uncertainty aversion, also in relation to financial consequences, are currently underestimated in water management (Pahl-Wostl 2007a). This needs urgent attention, as the risks and uncertainties connected to climate change can obtain disproportional attention, which may lead to substantial pressure from civil society towards inefficient and possibly ineffective measures.

1.4 EU RESEARCH POLICY RECOMMENDATIONS

This book discusses how European research has improved implementation of IWRM over the years and how these improvements relate to the current water management challenges. In this chapter we have shortly introduced Adaptive Water Management as the concept to improve water management in view of the increasing complexity and affects of climate change that constitute these challenges. We also discussed that a transition is needed towards adaptive water management regimes and the specifics which are anticipated to realise this transition. All this is further elaborated in Chapter 2. In this section we will wrap up the discussions and conclusions from this book into recommendations for further European research policy.

IWRM is based on the presumption that the complex socio-ecological nature of river basin environments and the inherent uncertainties associated with their management has to be taken into account in policy development and implementation. This integration has proceeded significantly in European research, but still additional efforts are needed, relating among others to the elements of social vulnerability like equity and fairness, and policy sectors like spatial planning. It is therefore recommended that the element of disciplinary integration be continued in European research.

The other basis of IWRM is that water management considers the entire catchment area as a functional unit. Although this element is acknowledged in European research, research and policy implementation still largely focuses on the national aspects. Economic analysis is for instance focused on national cost-benefits, where measures in a transboundary context could appear to be much more cost-effective. Another example is information management that usually deals with national information collection and processing, where collaboration across borders is necessary to support effective water management. Also legislation is mostly nationally focused; there are for instance only weak requirements for transboundary river basin districts in the EU-WFD. There is consequently a need for better cooperation and coordination across borders but such cooperation can be hindered by disputes over water. To achieve cooperation, at least formal arrangements and detailed procedures for transboundary cooperation must be established between riparian countries. European research should intensify its focus on building of trust and social capital for problem solving and collaborative governance in transboundary situations and at the same time research and develop theories on possibilities for cooperation in transboundary situations where cooperation is difficult because of a lack of trust.

One of the aspects of IWRM is the need for participation in a basin. Stakeholders and public participation are the keys to the successful implementation of water policies and design of transparent and open social learning processes is therefore a key requirement of sustainable water management regimes. A rich variety of public participation methods is available to engage the public. Little used but potentially attractive methods for public participation in water management next to hearings are public opinion surveys, are negotiated rule making, focus groups and citizens' juries. Research in this area should aim at participation in a transboundary context, taking into account that differences in water management competences and political processes of transboundary cooperation exist between countries.

IWRM seems to fall short in dealing with uncertainty and adaptive capacity of water management, and the concept of adaptive water management gets its main focus here. Uncertainty relates to the idea that selected water management strategies should be robust and perform well under a range of possible, but initially uncertain, future developments, more specifically when dealing with risk and vulnerability in relation to climate change. Adaptive capacity relates to the idea that water management has to trust that a collaborative process is a more robust strategy in conditions of uncertainty then any belief in prediction and control. However, entrenched perceptions and beliefs block innovation and change, and space has to be provided for creative and out-of-the-box thinking. European research should in this respect focus on successes and failures and the reasons behind these outcomes to build empirical evidence on approaches and the circumstances under which they perform best.

Finally, European research should pay ample attention to developing an integrated perspective on water information management, considering information as pervasive, supporting all the phases of the decision process. This relates to, among others, involving stakeholders in the information needs definition process, enhancing accessibility of the information also for non-expert people, integration of different sources of information, both technical and stakeholders-based, and adopting a "learning-by-doing" approach in monitoring system design. Through resolving these issues, improved information can be produced that improves adaptive decision-making under conditions of uncertainty.

In summary, from this book it is concluded that European water management research already addresses many of the future challenges. However, it is also concluded that to enable a transition towards adaptiveness in water management European research should focus specific attention towards:

- Improving disciplinary integration, among others on issues like equity and fairness, and policy sectors like spatial planning.
- Improving application of an integrative concept of vulnerability that includes resistance and resilience of systems next to exposure, and accounts for the human dimensions of vulnerability.
- Building of trust and social capital for problem solving and collaborative governance in water management.
- Developing governance models in transboundary situations where cooperation is difficult because of a lack of trust.
- Participation in a transboundary context, taking into account the differences in water management competences and political processes between countries.
- Building empirical evidence on successes and failures in water management strategies and the circumstances under which the strategies perform best.
- Developing an improved understanding of the consequences of climate change along with implementing new approaches to deal with uncertainty and scenario planning.
- Developing an integrated perspective on water information management in which information supports all the phases of the decision process.

All this should be based on creating a learning environment which enables close interaction between policy and science as well as between scientific disciplines. Furthermore, research should expand to experimenting with new multi-level governance approaches which combine bureaucratic hierarchies, networks and markets.

1.5 REFERENCES

Adger, W.N. (1999) Social vulnerability to climate change and extremes in coastal Vietnam. *World Development* **27**(2), 249-269.

Adger, W.N. (2006) Vulnerability. *Global Environmental Change* **16**, 268-281.

Allen, W., Bosch, O., Kilvington, M., Brown, I., and Harley, D. (2001) Monitoring and adaptive management: resolving social and organizational issues to improve information sharing in natural resource management. *Natural Resources Forum* **25**(3), 225-233.

Brooks, N. (2003) *Vulnerability, risk and adaptation: A conceptual framework.* Tyndall Centre Working Paper No.38

Buckingham Shum, S. (1997) Representing hard-to-formalise, contextualised, multidisciplinary, organisational knowledge. *AAAI Spring Symposium on Artificial Intelligence in Knowledge Management (Mar. 24-26, 1997).* Stanford University, AAAI Press, Palo Alto, CA, USA. http://kmi.open.ac.uk/people/sbs/org-knowledge/aikm97/sbs-paper2.html

Cash, D.W., Adger, W.N., Berkes, F., Garden, P., Lebel, L., Olsson, P., Pritchard, L., and Young, O. (2006) Scale and cross-scale dynamics: governance and information in a multilevel world. *Ecology and Society* **11**(2), 8 http://www.ecologyandsociety.org/vol11/iss2/art8/

Cosgrove, W.J. and Rijsberman, F.R. (2000) *World water vision; making water everybody's business.* Earthscan Publications Ltd. London, UK.

Dalton, J.R. (2004) *Democratic challenges, democratic choices. The erosion of political support in advanced industrial democracies.* Oxford University Press, Oxford, UK.

Downing, T.E., Aerts, J., Soussan, J., Barthelemy, O., Bharwani, S., Ionescu, C., Hinkel, J., Klein, R.J.T., Mata, L., Moss, S., Purkey, D., and Ziervogel, G. (2006) *Integrating Social Vulnerability into water management.* SEI Working Paper and NeWater Working Paper No. 4. Oxford: Stockholm Environment Institute

Falkenmark, M. (2000) *No freshwater security without major shift in thinking. Ten-year message from the Stockholm Water Symposia.* Stockholm International Water Institute, SIWI Stockholm, Sweden.

Folke, C., Hahn, T., and Norberg, J. (2005) Adaptive governance of social-ecological systems. *Annual Review of Environment and Resources* **30**, 441-473.

Giordano, R., Haase, D., Liersch, S., Timmerman, J.G., and Vurro, M. (2007) The role of information and new concepts for adaptive monitoring systems. *Ecology and Society* (submitted). www.ecologyandsociety.org

Gleick, P.H. (2003) Global Freshwater Resources: Soft-Path Solutions for the 21st Century. *Science* **302**, 524-528.

Gooch, G.D. (2004) The communication of scientific information in institutional contexts: The specific case of transboundary water management in Europe. Pages 13-29 in: Timmerman, J.G. and S. Langaas (eds.) *Environmental information in European transboundary water management*. IWA Publishing, London, UK. ISBN: 1 84339 038 8.

Gooch, G.D. and Stålnacke, P. (2006) *Integrated transboundary water management in theory and practice: Experiences from the new EU Eastern borders*. 1st, IWA Publishing, London, UK. ISBN: 1843390841.

Grossmann, M. (2005) *Kooperation an Afrikas internationalen Gewässern: die Bedeutung des Informationsaustauschs (Cooperation in Africa's transboundary waters: The role of information exchange)*. Discussion paper 9. Deutsches Institut für Entwicklungspolitik Bonn, germany.

GWP-TAC (2000) *Integrated Water Resources Management*. TAC Background Papers No. 4. Global Water Partnership - Technical Advisory Committee Stockholm, Sweden.

GWP-TEC (2003) *Poverty reduction and IWRM*. TEC Background papers no. 8. Global Water Partnership - Technical Committee Stockholm, Sweden.

Hinrichsen, D (2003) A human thirst. *World Watch magazine* January/February, 12-18 Worldwatch Institute, Washington DC, USA. www.worldwatch.org

Huitema, D., Egas, W., Moellenkamp, S., Mostert, E., Pahl-Wostl, C., and Yalcin, R. (2007) Adaptive water governance. Assessing adaptive management from a governance perspective. *Ecology and Society* (in press). www.ecologyandsociety.org

Ingram, H. (2001) Science and environmental policy. http://www.iec.uci.edu/S&EPpdf.pdf

IWA (2002) *Industry as a partner for sustainable development: Water management*. International Water Association and United Nations Environment Programme Beacon Press, London, UK.

Jønch-Clausen, T. (2004) *Integrated Water Resources Management (IWRM) and water efficiency plans by 2005. Why, What and How?* TEC background papers No. 10. Global Water Partnership Technical Committee Sweden.

Kabat, P. and van Schaik, H. (2003) *Climate changes the water rules: How water managers can cope with today's climate variability and tomorrow's climate change*. Synthesis Report of the International Dialogue on Water and Climate.

Klijn, E.H., van Bueren, E., and Koppejan, J.M.F. (2000) *Spelen met onzekerheid: over diffuse besluitvorming in beleidsnetwerken en mogelijkheden voor management*. Eburon, Delft, The Netherlands. 186 pp.

Krasner, S.D. (1983) *International regimes*. Cornell University Press, Ithaca, New York. 372 pp.

Lise, W., Timmerman, J.G., Vermaat, J., O'Riordan, T., Edwards, T., de Bruin, E., Kontogioanni, A., Barrett, K., Bresser, T., and Rochelle, E. (2004) Institutional and capacity requirements for implementation of the WFD. Pages 185-198 in: Vermaat, J. and others (eds.) *Managing European coasts: past, present and future*. Springer Verlag, ISBN: 3-540-23454-3.

Mayer, P., Rittberger, V., and Zürn, M. (1993) Regime Theory: State of the art and perspectives. Pages 391-430 in: Rittberger, V. and P. Mayer (eds.) *Regime Theory and International Relations*. Clarendon Press, Oxford.

Mayntz, R. (1998) *New challenges to governance theory*. Jean Monet Chair Papers No. 50. European University Institute

Mayntz, R. (2003) From government to governance: Political steering in modern societies. From Government to Governance: The Case of Integrated Product Policy (IPP), Würzburg, Germany. http://www.ioew.de/governance/english/veranstaltungen/Summer_Academies/SuA2Mayntz.pdf

Medema, W., McIntosh, B.S., and Jeffrey, P. (2007) Observations on fulfilling the promise of the IWRM and Adaptive Management concepts. *Ecology and Society* (in prep.).

Merrey, D.J., Drechsel, P., Penning de Vries, F.W.T., and Sally, H. (2005) Integrating "livelihoods" into integrated water resources management: taking the integration paradigm to its logical next step for developing countries. *Regional Environmental Change* **5**(4), 197-204. http://www.springer.com/sgw/cda/frontpage/0,,5-40532-70-1136209-0,00.html

Moberg, F. and Galaz, V. (2005) *Resilience: Going from conventional to adaptive freshwater management for human and ecosystem compatibility*. Swedish Water House Policy Brief Nr. 3. SIWI, Stockholm, Sweden

Mostert, E. (2005) *How can international donors promote transboundary water management?* Discussion paper 8. Deutsches Institut für Entwicklungspolitik Bonne, Germany.

Nilsson, S. (2003) The role and use of information in transboundary water management. Kungl Tekniska Högskolan, Department of Land and Water Resources Engineering, http://www.lwr.kth.se/Publikationer/PDF_Files/LWR_LIC_2012.pdf

OECD (2001) *Engaging citizens in policy-making: Information, consultation and public participation*. PUMA Policy brief No. 10. Organisation for Economic Cooperation and Development Paris, France.

Pahl-Wostl, C. (2002) Participative and stakeholder-based policy design, evaluation and modeling processes. *Integrated Assessment* **3**(1), 3-14.

Pahl-Wostl, C. (2007a) The implications of complexity for integrated resources management. *Environmental Modelling and Software* **22**, 561-569.

Pahl-Wostl, C. (2007b) Transition towards adaptive management of water facing climate and global change. *Water Resources Management* **21**(1), 49-62.

Pahl-Wostl, C., Downing, T., Kabat, P., Magnuszewski, P., Meigh, J., Schlueter, M., Sendzimir, J., and Werners, S. (2005) *Transition to Adaptive Water Management; The NeWater project*. NeWater Working Paper 1. Institute of Environmental Systems Research, University of Osnabrück Osnabrück, Germany.

Pahl-Wostl, C. and Hare, M. (2004) Processes of social learning in integrated resources management. *Journal of Community & Applied Social Psychology* **14**(3), 193-206.

Perrings, C. (2005) Environmental risk, uncertainty and the precautionary principle. Pages 37-46 in: Koeman, J.H. and J.D. Schiereck (eds.). *Responsibilities of Environmental Research*. Royal Netherlands Academy of Arts and Sciences, Amsterdam, The Netherlands. http://www.knaw.nl/publicaties/pdf/20051011.pdf

Raadgever, G.T., Mostert, E., Kranz, N., Interwies, E., and Timmerman, J.G. (2007) Adaptive management of transboundary river basins - analysis of transboundary regimes from a normative perspective. *Ecology and Society* (submitted). www.ecologyandsociety.org

Rauch, W. (1998) Problems of decision making for a sustainable development. *Water Science and Technology* **38**(11), 31-39.

Rhodes, R.A.W. (1997) *Understanding Governance: Policy Networks, Governance, Reflexivity, and Accountability*. Buckingham, Philadephia: Open University Press,

Ridder, D., Mostert, E., and Wolters, H.A. (2005) *Learning together to manage together - Improving participation in water management*. ISBN 3-00-016970-9. Druckhaus Bergmann Osnabrück, Germany.

Saeijs, H.L.F. and van Berkel, M.J. (1995) Global water crisis: the major issue of the 21st century, a growing and explosive problem. *European Water Pollution Control* **5**(4), 26-40.

Savenije, H.G. and van der Zaag, P. (2000) Conceptual framework for the management of shared river basins; with special reference to the SADC and EU. *Water Policy* **2**, 9-45.

Schulze, G.G. and Ursprung, H.W. (2001) Introduction. Pages 1-14 in: Schulze, G.G. and H.W. Ursprung (eds.) *International Environmental Economics: A Survey of the Issues*. Oxford University Press, ISBN: 0199261113. 301 pp.

Serageldin, I. (1994) Making development sustainable. Pages 1-3 in: Serageldin, I. and A. Steer (eds.) *Making development sustainable: from concepts to actions*. Environmentally Sustainable Development Occasional Paper Series The World Bank, Washington D.C., USA.

Smouts, M.-C. (1998) The proper use of governance in international relations. *International Social Science Journal* **50**(1), 81-89.

Timmerman, J.G. and Langaas, S. (2004a) Conclusions. Pages 240-246 in: Timmerman, J.G. and S. Langaas (eds.) *Environmental information in European transboundary water management*. IWA Publishing, London, UK. ISBN: 1 84339 038 8.

Timmerman, J.G. and Langaas, S. (2004b) *Environmental information in European transboundary water management*. IWA Publishing, London, UK. ISBN: 1 84339 038 8.

Timmerman, J.G. and Langaas, S. (2005) Water information - what is it good for? On the use of information in transboundary water management. *Regional Environmental Change* **5**(4), 177-187. http://www.springer.com/sgw/cda/frontpage/0,,5-40532-70-1136209-0,00.html

Timmerman, J.G., Ottens, J.J., and Ward, R.C. (2000) The information cycle as a framework for defining information goals for water-quality monitoring. *Environmental Management* **25** (3), 229-239.

van der Brugge, R., Rotmans, J., and Loorbach, D. (2004) The transition in Dutch water management. *Regional Environmental Change* **5**(4), 164-176. http://www.springer.com/sgw/cda/frontpage/0,,5-40532-70-1136209-0,00.html

Wolf, A.T., Yoffe, S.B., and Giordano, M. (2003) International waters: identifying basins at risk. *Water Policy* **5**, 29-60.

World Bank (2002) *Linking poverty reduction and environmental management - Policy challenges and opportunities*. DFID, EC, UNDP, and World Bank Washington, USA.

2

New approaches to water management: contributions from European research on complexity, learning and uncertainty

Claudia Pahl-Wostl and Paul Jeffrey

2.1 INTRODUCTION

Over the past decade, a major change in the rhetoric surrounding water resources management has been evident. The debate is now dominated by an increased awareness of integrated management approaches taking into account environmental, economic and social considerations and by the search for solutions, which go beyond technical end-of-pipe solutions. The importance of improving water governance is now widely recognised.

For the past two decades, new, more integrated approaches have been developed and are currently implemented to address perceived shortcomings. During the last decade, the principle of IWRM (Integrated Water Resources Management) has been an attempt to implement such integrated approaches in water management (GWP-TAC 2000). "Integrated" clearly indicates an aspiration to functionally engage a range of perspectives by taking all potential trade-offs at different scales in space and time into account. Such an approach attempts to overcome the limitations of technical, end-of-pipe solutions dealing with individual problems in isolation and thus often neglecting unexpected consequences (Pahl-Wostl 2007a).

The European Water Framework Directive (WFD) is a major piece of European legislation which signals a change in water policy towards more integration and participation. The WFD has a holistic goal of achieving a good state of all European Waters by 2015 based on the development of River Basin Management Plans. It prescribes the involvement of all interested parties and the general public in developing, implementing and updating such plans. In a similar spirit, the European Water Initiative launched during the Sustainability Summit in Johannesburg 2002 aims at supporting developing countries to implement IWRM Plans by 2020 (http://www.euwi.net/).

One can summarise that the need to move towards integrated approaches, to involve a wide range of stakeholders and to foster learning and change can be explained by several insights that are not entirely new but have only recently been seriously taken into account in management practice:

- Increasing interdependence between government bodies and other stakeholders (e.g. collective decisions, distinctive competences and complementary contributions needed) reduces the efficacy of a traditional command and control management style and requires a shift towards a more interactive and participatory style.
- Increasing complexity of natural resources management (e.g. a shift towards integrated approaches in management objectives, integrated solutions instead of technical fixes, increased awareness for the complex nature of socio-ecological systems) requires an enhanced capacity for learning and innovation and the need to involve a wider group of stakeholders.

- Increasing uncertainties (e.g. climate change, dynamic socio-economic conditions) require a more adaptive and flexible management approach to realise a faster coping cycle, which allows rapid assessments of, and implementation of the consequences of new insights. This requires new skills and capabilities, informal management structures and the inclusion of expert knowledge as well as local lay knowledge.

Such changes in perception and regulatory frameworks imply a need to take into account the full complexity of the systems to be managed. At the same time one observes an increase in uncertainties due to climate change and globalisation. In particular, in developing and threshold countries, socio-economic change proceeds at an unparalleled fast pace. The established planning approaches in water management developed in industrialised countries to cope with specific problems such as floods or pollution relies strongly on the ability to predict the effect of management measures and to design systems that can be controlled. Such approaches are less appropriate in highly uncertain and complex situations, in particular if more integration is one management objective.

However, progress in developing, and in particular in implementing, innovative management approaches has not kept pace with the expressed need for a change in management paradigms and practices. The gap between political rhetoric and operational implementation remains large. Integrated and adaptive management approaches are required that perform well under complex and unpredictable conditions and that can be tailored to the institutional, cultural, environmental, and technological settings of river basins. We explore the principle contending approaches in this field in the following section.

2.2 CONTEMPORARY APPROACHES TO WATER MANAGEMENT

The past 30 years have seen radical changes in the styles of thinking about (a) relationships between society and the environment (see Buttel 1986), (b) the nature of utility in the context of natural resources management (Georgescu-Roegen 1982) and (c) desirable outcomes from human interventions in natural processes (Selfa 2004). Increasing awareness that natural capital requires protection, that communities should be guiding their own development, and that the Newtonian model is constrained in its ability to inform complex problems, have motivated new approaches to managing natural resources. In the field of water management, two strategic level models have become particularly influential; Integrated Water Resources Management and Adaptive Management; both of which claim to promote sustainable use of the resource.

IWRM has been advocated as the most sustainable means to incorporate the multiple competing and conflicting uses of water resources ever since the first UNESCO International Conference on Water, which took place in 1977 at Mar del Plata, Argentina. The Global Water Partnership (GWP) defines IWRM as '*a process which promotes the co-ordinated development and management of water, land and related resources in order to maximise the resultant economic and social welfare in an equitable manner without compromising the sustainability of vital ecosystems.*' (GWP-TAC 2000). Whilst there have been suggestions for modifications to this designation (e.g. Jonker 2002 p719), the GWP version remains the most oft quoted version. As an ambition, IWRM therefore seeks to address (simultaneously!) two highly complicated and complex problem sets; sustainable development and cross sectoral planning. Recently various authors have argued that the IWRM approach, as it is defined by the GWP, cannot be implemented in practice generally due to operational questions and related problems of establishing measurable criteria (e.g. Jeffrey and Gearey 2006; Kluge 2005; Lankford and Cour 2005; Biswas, 2004).

Biswas (2004) and several respondents to his article pointed out a number of barriers to implementation. Integration of sectors and issues would require more centralised policy development and implementation and thus larger, slower, and more bureaucratic authorities to handle all policy aspects. Furthermore, objectives like stakeholder participation and decentralisation would be unlikely to promote integration. In summary the following main limitations of IWRM have been highlighted in these discussions:

- Vagueness of the concept,
- Integration of many topics and management of a system for a specific purpose cannot be realised simultaneously,
- The claims for change in management practice made by the promoters of IWRM seem to be based on normative claims rather than a sound scientific base that would provide evidence of these new management approaches,
- Unambiguous evidence for the successful implementation of IWRM is lacking,
- An emphasis on process without clearly defined and measurable targets for the goals to be achieved.

These serious claims suggest strongly that more emphasis needs to be given to share lessons learned and to identify key barriers and drivers for the implementation of integrated and sustainable water management approaches.

The origin of the Adaptive Management concept lies in many different intellectual and practitioner fields, but its initial presentation as a natural resources management paradigm was in the 1970s (NRC 2004). It was initially developed at the University of British Columbia in Canada and at the International Institute for Applied Systems

Analysis in Vienna to support the management of natural resources under uncertainty and then further promoted by an international network, the resilience alliance (Holling 1978; Walters 1986; Walters and Holling 1990; Irwin and Wigley 1993; Parma *et al.* 1998; Ohlson 1999; Prato 2003). The AM concept has been widely presented as the approach which should be adopted by natural resource managers, building on the recognition that ecosystems are complex systems, which are 'adaptive' or 'self organising' and that management systems must be able to readjust to change or surprise in the system being managed (Gunderson and Holling 2001). AM has been described as an integrated, multidisciplinary and systematic approach to improving management and accommodating change by learning from the outcomes of management policies and practices (Holling 1978).

AM as process includes the design and implementation of management programs that offer the possibility to experiment with and compare selected policies and practices. This comparison takes place through evaluation of alternative hypotheses about the system (Holling 1978; Walters 1986; Lee 1999). Lee (1993) emphasises the usefulness of this approach by stating: 'if human understanding of nature is imperfect, then human interactions with nature should be experimental'. Although Holling (1978) is of the opinion that the AM concept will support management to proceed even if there is a lack of sound scientific foundation for action, other authors contradict this by suggesting that AM requires major investment in research, monitoring, and modelling to test alternative hypotheses about sustainable use and management of the natural resources (Smith and Walters 1981; Hilborn *et al.* 1995; Walters and Green 1997; Prato 2003).

In this light AM could be seen as a management process that is both anticipatory as well as adaptive (Kay 1997). Hypothesis and assumptions are developed based on a thorough understanding of the system as a way to anticipate to possibilities and uncertainties that could have an impact on this system. These hypothesis and assumptions are translated into plans and actions, which are evaluated and monitored in order to test their affect on the system. Based on these results, the hypotheses and assumptions are adapted with the objective of improving the overall management framework. The idea is that this process is repeated to guarantee continuous improvement.

Both of these management philosophies promise significant advantages over previous approaches to managing water. Indeed, they aspire to deliver balanced, long-term solutions, which meet the demands of social equity and sustainable development. This sounds indeed like an idyllic future. But do we have the opportunity to implement such plans? In Europe, we do, in the form of the Water Framework Directive.

2.2.1 The European Water Framework Directive – an opportunity to apply novel water management strategies?

In principle the WFD provides a flexible framework for implementation that would allow innovative, participatory and adaptive approaches to water management. However, the current implementation process of the WFD seems to follow largely a traditional command and control approach where operational targets are defined to be able to "quantify" the good ecological state of water bodies – e.g. European wide binding water quality targets. One may argue if this approach will really lead to the goals envisaged by the WFD. In the following some issues are addressed that highlight the challenges of this new water policy for water management regimes that have largely been shaped by a control approach to policy.

The European Water Framework Directive is novel in that it leaves far more opportunities for member states to adapt the implementation of the Directive to national and regional specificities. Whereas the time schedule is narrowly prescribed, the relative vagueness of the goal of a "good ecological status" leaves a lot of freedom for interpretation.

The European legislator should not reasonably prescribe every detail of implementation – particularly not within a 'framework' (sic!) directive, but leave many decisions regarding the operationalisation of goals and measures to the member states and the river basin authorities (Newig *et al.* 2005). The rationale of this applied subsidiarity principle (cf. Collier 1997; Kraemer 1998) is, of course, that despite the overall goal to reach good water status, regional conditions differ in terms of culture, water uses, water-related problems and so forth. However, the openness of the directive is also partly a product of watered-down compromises (Kaika 2003, p. 311). In any case, it is both sensible and requisite that a considerable amount of decisions are made on a regional level by authorities familiar with regional conditions. This extended discretionary scope on the part of the authorities entails normative uncertainty (Arentsen *et al.* 2000), i.e. normative considerations of goals and tastes (cf. March 1978). Such uncertainty about the final targets to be achieved may be perceived as a problem for effective and efficient implementation (inability to act prior to having agreed on binding targets) or as an opportunity to negotiate goals that can really be achieved.

In his analysis whether the Framework Directive (WFD) builds resilience, i.e. the capacity of freshwater systems to deal with change and perturbations, Galaz (2005) concluded that the current realisation of the WFD in Sweden raises some considerable issues, and might at worst reduce the resilience of nested social-ecological freshwater systems. In particular he highlighted that:

- Collective action and analytical deliberation is highly limited
- Water management institutions disregard complexity and uncertainty
- Water policy is poorly prepared to tackle global environmental change

Such insights are not limited to Sweden. Despite the stated policy goals by government to foster innovation and more flexible and adaptive management approaches, one observes a lack of change at the operational level. In the Netherlands, for example, the government asks for a radical rethinking of water management – more space for rivers and living with water rather than control. However, management practice is very slow in adopting new strategies. Again the barrier for adopting more integrated and adaptive approaches is not the guiding policy framework but rather the approach how it is implemented in operational management. Currently, adaptive management seems to be limited in prevailing designs, practices and ideas surrounding river basin management (Pahl-Wostl *et al.* 2006; Pahl-Wostl, 2007b).

Consequently, we now take the identified areas of concern as an agenda for commenting on the contributions of four EU funded research activities towards the development and application of IWRM and AM approaches. Specifically, we consider the extent to which such approaches are able to address;

- Uncertainties and their role in policy and management
- Complexity and dynamics of human-technology-environment systems
- Learning and managing change

2.3 CONTRIBUTIONS FROM DIFFERENT EU FUNDED PROJECTS

Numerous research and demonstration activities funded by the EU have addressed, either directly or in passing, the theory and practice of IWRM. Hardly any have specifically considered Adaptive Management. The four projects that form a focus for our commentary in this contribution address one or all of the above-mentioned themes. These four projects (Tables 1 & 2) have been selected based on a survey conducted among a wide range of water related projects funded by the EU framework programmes. The chosen projects provide a representative overview of how complexity, learning and uncertainty are addressed and understood in EU funded research on catchment management. The list is by no means exhaustive but we have chosen four projects which encompass a variety of planning and management components and which adequately represent the major features of contemporary thinking on catchment management. One can note that the three projects funded under FP5 focus mainly on one of the key areas of interest whereas the NeWater project aims at addressing all. This can be interpreted as a change in thinking and emphasis in water management research.

Table 1: Some characteristics of the projects

Topic	Project			
	HarmoniRiB	**AQUADAPT**	**HarmoniCOP**	**NeWater**
EC project reference #	EVK1-CT-2002-00109	EVK1-CT-2001-00104	EVK1-CT-2002-00120	GOCE-CT-2003-511179
Duration (month/year)	10/02 – 09/06	02/02 – 02/05	10/02 – 12/05	01/05 – 12/08
Number of participating organisations	10	13	17	36
Case studies – regions, river basins	8 Europe	5 Europe	10 Europe	4 Europe, 2 Africa, 1 Central Asia
Degree of balance between natural and social sciences	Natural science focus	Balanced	Social science focus	Balanced
Involvement of stakeholders	low	medium	high	high
Uncertainties	+++	+	+	+++
Evolutionary Integrated System Dynamics	+	+++	-	+++
Learning and managing change	-	+	+++	+++

Table 2: The scale of stakeholder participation within the project

Stakeholder Group	HarmoniRiB	Aquadapt	HarmoniCOP	NeWater
Citizens	0	3	0	0
NGO's	0	2	3	3
Companies & consultants	2	3	3	3
Local government (operational)	0	1	2	3
Local government (policy)	0	3	3	2
Regional government (operational)	2	1	2	3
Regional government (policy)	0	3	3	3
National government (operational)	0	0	2	3
National government (policy)	1	1	3	2
Continental (e.g. EU)	0	2	1	2
Water authorities	2	3	2	3
Other 1, namely farmers	0	0	0	3
Other 2, namely irrigation associations	0	0	2	3
Poor communities (community leader)	0	0	0	3

Scoring: 0: Not involved; 1: Informed (one-way communication); 2: Consulted (limited two-way communication); 3: Active participation (active discussion throughout the project)

2.3.1 HarmoniRiB

The overall goal of HarmoniRiB was to develop methodologies for quantifying uncertainty and its propagation from raw data to concise management information (Reefsgaard *et al.* 2005).

A comprehensive approach to classify uncertainties was developed. The perspective on the policy process in which this information is embedded was strongly informed by formal decision theoretical approaches. The concepts developed looked at the framing of the initial decision making process, the prediction of effects of measures and the economic assessment of costs and benefits. Uncertainties in data and in models were explored in depth. The main specific outputs from HarmoniRiB were:

(1) Uncertainty concepts and tools
- Development of the Data Uncertainty Engine (DUE) software tool and manuals
- A report with guidelines on how to assess uncertainty in various types of data
- Guidelines for supporting incorporation of socio-economic uncertainties in the decision making process
- A report describing uncertainties in the economic analysis of the WFD.

(2) Databases for accommodating uncertain data
- Database design, database software and associated manuals.

(3) River basin network and data
- Datasets from eight river basins including the data uncertainty of the data and the documentation of the data and the associated uncertainty as uploaded to the HarmoniRiB Data Centre.

(4) Case studies
- Reports from eight case studies comprising test of concepts for HarmoniRiB methodologies and tools on cases focussing on different aspects of the WFD.

The emphasis of HarmoniRiB was clearly to develop a sound approach to make uncertainty in decision-making processes explicit and quantify them if possible. The importance of adaptive management and learning cycles has not been addressed.

The case studies showed that assessing uncertainty in data may in many cases be a complex and challenging task. When uncertain data are fed into models with uncertainties related to parameter values and model structures (process equations) the uncertainty of model predictions become very complex and may be difficult to comprehend for water managers, stakeholders and other non experts. Another key conclusion from the case studies was that framing of the decision problem and externally given boundary conditions (such as climate change and regulatory policy decisions) often are difficult to present in an easily understandable manner together with uncertainty originating purely from data and models. In some cases the uncertainties related to framing, context and economical aspects were found to be more important than data and model uncertainty itself.

The findings strongly suggest that more emphasis needs to be given to analysing the processes of framing preceding decision making and to develop approaches that overcome the limitations of formalised decision theoretical approaches.

2.3.2 AQUADAPT

The need to place strategic water management for arid and semi-arid areas on a robust theoretic footing has presented a challenge to the scientific and practitioner communities for several decades. Meaningful debate on the relative merits of different water policy mechanisms can only be conducted in the context of a broader appreciation of how water supports and nurtures our societies. The starting point for the AQUADAPT project has been a perspective on this relationship, which emphasises connectivity and feedback. The challenges facing those charged with planning and managing our water resources are increasingly characterised by inter-scale, multiple-element and dynamic patterns of demand and supply. That the various aspects of the water environment (water availability & quality, land use, governance regimes, etc.) are mutually influential is therefore an intuitively powerful concept. The ambition of the project has been to explore just how useful this concept might be.

A set of insights has been generated by work on complex, co-dynamic and co-evolutionary processes, some of it critical of the use of the coevolutionary analogy (see Winder *et al.* (2005) and Jeffrey & McIntosh (2007) for relevant discussions generated through the Aquadapt project). So how can a complexity-embracing, co-dynamic model of change contribute to questions of water management policy? The first point to emphasise here is that such an understanding allows a better "positive" depiction of complex dynamics and thus can act as a basis for a more informed decision and intervention. Secondly, interventions designed distant, hierarchically or spatially, from the intended level of effect are likely to fail as it is more difficult (or comes with higher "transaction" costs) to recognise and adapt to local conditions. Indifferent as they are to the diversity of local behaviours (e.g. EU policies versus diverse regional socio-cultural realities) they are likely to be ineffective and not "selected". Additionally, given the limited possibility for prediction of a change path, massive interventions are inherently more risky. Multiple small-scale policy "experiments" where sub-elements may be "selected" according to local conditions, followed by close continuous monitoring and process revision following implementation are preferable in this respect. Theories of Adaptive Management may indeed be useful in this context.

On the other hand, a co-dynamic view recognises the importance of large-scale events and systems in relation to lower-level ones and the potential of a macro-micro level gridlock. For example, development policies may continue to favour large infrastructure based solutions to water stress despite evidence of in-effectiveness because of a lock-in of bureaucracies and policies at institutional levels and as conditioned by the overall political environment. It is thus some proper combination of diverse intervention at different hierarchical/spatial scales, which is preferable from a co-dynamic perspective.

The preference for small-scale, diverse interventions is supported by insights from studies in synergetics and self-organisation/chaotic mathematical models. These emphasise how small changes, which are spatially and temporally resonant to system dynamics events, rather than large energetic changes, are the ones that cause large results. In evolutionary terms "resonant" interventions can be likened to those at an early stage of a coevolutionary path-dependent track, e.g. when a new technology first emerges or after a "shock" where key conditions are "locked-out" and redefined. In technology policies for example, it might be relevant to structure a strategy of diverse investment and regulatory support of several technologies at an early stage of development of a new technology to reduce the possibility of an early "lock-in" of an inefficient alternative.

Finally, a co-evolutionary (as distinct from a co-dynamic) perspective suggests that if there is something close to a dominant selection criterion in the "survival" of "populations" (policies, firms, etc) this is not efficiency, but "learning", i.e. the ability to continuously adapt to the co-changing and at times unexpectedly and dramatically altered, conditions. Again it would be wrong to interpret this as a call to an "incremental" mode of policy-making and a return to ad hoc, short-term corrective decisions. Nor as a justification of "business as usual" behaviour, i.e. waiting for "adaptation" when the time to face the consequences comes as the adaptive coevolutionary path at that time might turn out to be catastrophic. Instead, coevolution balances between stressing the limits to rationalism and forecasting vs. the need for creating and working towards (uncertain) visions of the future.

Flexible planning approaches, such as the Strategic Niche Management approach, fit this perspective well. In terms of policy analysis, studies such as socio-technical scenarios or visions, where alternative "futures" are explored fit much better with the philosophy of coevolution than conventional "policy assessment" or "policy evaluation" models. In terms of ex-post policy evaluation, a coevolutionary perspective stresses that past policies should be understood within their coevolutionary context and therefore on the contingencies of their time. Indeed, a coevolutionary perspective suggests that all past actions are in some sense "right" as they are the precedent of a coevolutionary path. Again, this does not imply an a-priory acceptance of what has happened. On the one hand, one could assess the values with which policies emerged and whether these were indeed the values primarily intended by the policies. On the other hand, one could impose present day values and see how past policies fit with them.

The Aquadapt project's contribution to the debate on learning is less expansive. There is evidence from the project that European citizens lack knowledge of the water cycle, in particular about the relationship between natural and engineered systems. The project also identified a lack of interest in participating in public debates about water management. Findings from studies of governance arrangements across Europe suggest that whilst there is already a long tradition of adaptation in the water management sector, existing modes of water management and water utilisation are often preserved due to the fact that property rights do matter and maintain rivalries. Where adaptive responses can be identified they are often triggered by a sense of urgency. However, significant changes to the structure or terms of reference of governance bodies are more likely to be initiated by broad political initiatives rather than as a response to disparity in the supply-demand balance.

Uncertainty was addressed within the Aquadapt project largely through its detailed evaluation of the suitability and utility of a coevolutionary perspective on water management. The role that uncertainty plays in determining appropriate approaches to sustainable development is crucial here. It would be nonsensical to propose that solving current problems will assure sustainability for all time. New problems and threats will emerge, variations of previously addressed problems will arise, and today's problems will evolve new dimensions. Given mankind's dismal record of forecasting the future, we can assume that we have little idea what form future threats may take. The specifics of tomorrow's challenges are thereby both unknown and largely unknowable. It is for this reason (the inherent and unavoidable uncertainty of the future) that there is a need to seek out knowledge concerning sustainability that is more generic in nature i.e. societal strategies that are effective across a range of temporal and spatial scales. As '*we cannot.....guarantee the persistence of any particular system in perpetuity' we are led to a view of sustainability as a 'normative ethical principle' which has 'no single version' and is 'a process, not a state.*' (Robinson *et al.* 1990). An interesting corollary of this perspective is that a focus on the development of collective action to "lock-out" undesired effects misses the question, which is what does constitute an "undesired" effect and furthermore how different alternative actions to effect a desired lock-out can be compared.

2.3.3 HarmoniCOP

The main objective of the HarmoniCOP (Harmonizing COllaborative Planning) project was to increase the understanding of participatory river basin management in Europe and the role of IC tools in river basin management. The project adopted a social learning perspective to improve the scientific understanding for the role of participatory processes in water management and to develop practical guidance for the design of participatory processes and the implementation of the European Water Framework Directive.

Box 1 – What is social learning?

> Social learning in river basin management refers to developing and sustaining the capacity of different authorities, experts, interest groups and the public to manage their river basins effectively. Collective action and the resolution of conflicts require that people recognise their interdependence and their differences and learn to deal with them constructively. The different groups need to learn and increase their awareness about their biophysical environment and about the complexity of social interactions.
>
> Why is social learning needed to move towards and to sustain integrated, adaptive water management? As pointed out in the previous section, technical infrastructure (e.g. large technical infrastructure for flood protection), citizen behaviour (expectations regarding safety in floodplains, risk perception) and habits, and engineering rules of good practice are often mutually dependent and stabilise each other with the effect that changes towards new and improved resource management schemes are blocked (Pahl-Wostl 2002). Social learning is assumed to be crucial to break up such "lock-in" situations. It is required for implementing change towards and for sustaining adaptive management practices.

Participatory management and stakeholder involvement have gained increasing importance in water resources management (ICWE 1992; Global Water Partnership 2000; Mostert 2003; Bouwen and Tailleu 2004; Pahl-Wostl 2002, 2007a). In integrated resource management one has often to deal with undefined problem situations where different perspectives on a problem exist. Due to the complexity of the systems under consideration and the presence of high uncertainties more than one interpretation may be equally plausible. Hence rather than finding an optimal and best solution one has to design a social learning process to develop an understanding for and negotiate about different legitimate perspectives.

A new concept for social learning in river basin management has been developed in the context of the European project HarmoniCOP. Figure 1 represents the framework for social learning developed to account for learning processes in water resources management (Craps 2003; Pahl-Wostl 2002; Pahl-Wostl *et al.* in press). The framework is structured into context, process and outcomes and a feedback loop to account for change in a cyclic and iterative process. The context refers the governance structure and the natural environment in a river basin. To improve the state of the environment implies in practice implies most often a change in the governance structure.

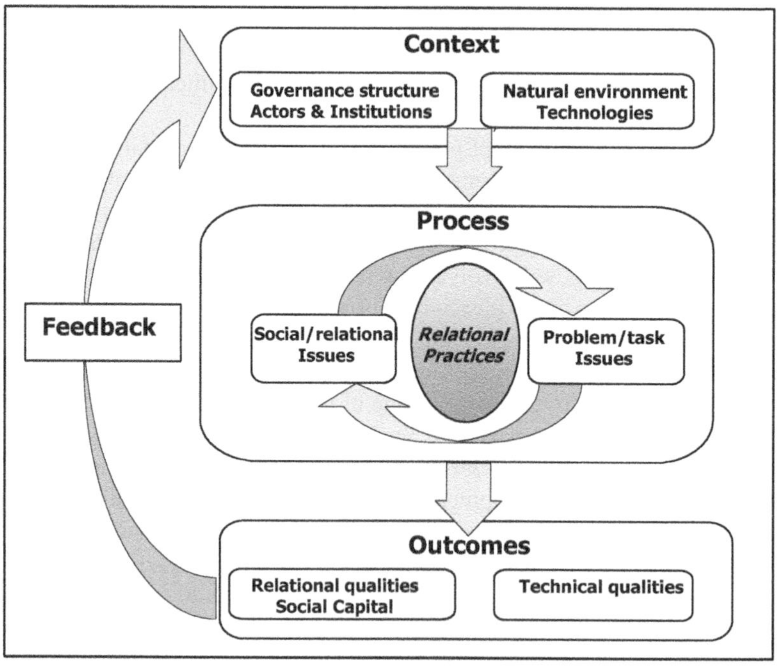

Figure 1: Conceptual framework for social learning in resources management. In the centre are multi-party processes that are influenced by the context in which they are embedded and that produce outcomes that may lead to changes in the context and thus to a cyclic and iterative long-term process of change.

The process concept referring to multi-party interactions in actor networks has two pillars (Figure 1). They relate to the processing of factual information on a problem (content management) and engaging in processes of social exchange (social involvement). Social involvement refers to essential elements of social processes such as the framing of the problem, the management of the boundaries between different stakeholder groups, the type of ground rules and negotiation strategies chosen or the role of leadership in the process.

In particular during the initial stages of dealing with a problem, the framing and reframing of a problem domain determine the direction of the overall process. Actors hold frames that determine how they give sense and meaning to information and their physical and social environment. Frames may derive from culture, social role, scientific disciplines etc. Differences in the framing of an issue are among the key reasons for problems in communication and entrenched conflicts among actors. The framing of an issue includes for example what is at stake and who should be included and in which role. Processes of framing and reframing are essential elements of social dynamics in a group during processes of negotiation of meaning. Powerful actors often impose their frame, their interpretation of an issue, onto a process. A relational practice may be a moderated role playing game or policy exercise where actors are willing to reflect and discuss about their own perspectives and listen to others. Social learning must not lead to consensus but to develop the ability to deal with differences constructively.

The overall social learning process in a group leads to both technical qualities such as the improvement of the state of the environment, and to relational qualities and social capital such as an increase in the capacity of a stakeholder group to manage a problem.

Table 3 summarises results from the case studies in HarmoniCOP regarding factors that constrain and that support social learning (Tippet *et al.* 2005; Mostert *et al.* 2007).

Table 3: Factors that constrain and support social learning.

Factors constraining SL	***Factors supporting SL***
STRUCTURE - CONTEXT	***STRUCTURE - CONTEXT***
o Centralised political and economic systems o Privatisation and commercialisation of environment. o Rigid bureaucratic systems. o Political secrecy and poor public access to information.	o Increased decentralisation of power o Move away from overregulated bureaucracy o Political recognition of the positive value of the public voice o Greater environmental awareness by members of the public o Developing a more consensus based culture
PROCESS	***PROCESS***
o Lack of clear objectives & process for involvement o Lack of time and effort taken to build trust o Lack of process to explore common ground rules and manage conflicts constructively o Lack of process to link planning at different levels of scale o Ineffective communication of technical issues	o Provision of sufficient time and resources o Opportunities for participation early enough in process o Use of facilitators and process management o Definition of commonly accepted ground rules o Explicit recognition of different perspectives

The HarmoniCOP project has provided some evidence that processes of social learning and the presence of informal actor platforms are of major importance to implement and sustain integrated and sustainable water resource management. However, it is of crucial importance to better understand the interplay between formal and informal institutions and the role of bridging organisations.

Important implications for water governance and water management:

(1) The principle of integrated water resources management has been criticised as being unrealistic. This critique was described above in Section 2.2. However, the HarmoniCOP project portrays another perspective on how integration can be achieved – that of a more dynamic actor landscape where integration is not achieved by bureaucratic hierarchies but rather by processes of network governance. It also highlights the need and the requirements for processes of social learning to build the capacity to come to joint solutions and to make thus stakeholder participation effective in terms of achieving the goals of water management.

(2) Water management is facing increasing uncertainties due to climate change, fast changing socio-economic boundary conditions and the goal of integration over a wider range of objectives. As a consequence, effective water governance must be adaptive. The paper highlights structural elements and processes in water governance regimes that increase adaptiveness without compromising stability.

(3) In most countries the structural conditions for integrated and adaptive water management are not yet given. Hence there is a need for major changes where the kind of learning processes investigated in the HarmoniCOP project will most likely play a major role.

Challenges still lie ahead, in particular in further collecting empirical evidence and in using available empirical studies for comparative analyses. The results from HarmoniCOP suggest that the development of such adaptive institutional settings involves continued processes of social learning where stakeholders at different scales are connected in flexible networks and where sufficient social capital and trust is developed to collaborate in a wide range of formal and informal relationships from formal legal structures and contracts to informal, voluntary agreements.

2.3.4 NeWater

The project NeWater (New Approaches for the Adaptive Water Management Under Uncertainty) develops new methods for integrated water management taking into account the complexity of the river basins to be managed and the difficulty in predicting the factors influencing them, e.g., climate, socioeconomic developments. NeWater focuses, in particular, on the transition from current regimes of water management in a river basin to more integrated, adaptive approaches. It embraces all aspects that have been identified to be of major concern, namely the nature and role of uncertainties, complex dynamics of human-technology-environment systems and learning and managing change.

NeWater uses a broad definition of adaptive management as a systematic process for improving management policies and practices by learning from the outcomes of implemented management strategies. The whole adaptive management process requires a number of steps that are part of an iterative policy cycle:

- In the definition of the problem different perspectives need to be taken into account in a participatory process.

- The design of policies should include scenario analyses to find strategies that perform well under different possible but initially uncertain future developments and to identify key uncertainties.
- Decisions should be evaluated by the costs of reversing them.
- The design of monitoring programmes should include different kinds of knowledge to become aware of undesirable developments at an early stage.
- The management cycle must include institutional settings where actors assess the performance of management strategies and implement change if needed in a transparent way.

The implementation of such a management approach is only possible if certain structural conditions are fulfilled. Hence the implementation of adaptive management needs integrated system design. In the past water management regimes have been designed to be controllable. A changed perspective aims at endowing water management regimes with properties of complex adaptive systems where control is distributed and adaptation to changing conditions possible.

NeWater has developed a set of assumptions on the features needed for a management regime to be adaptive and to maintain environmental, economic and social sustainability in a fast changing and uncertain world. Two different regimes are contrasted as the extreme, opposing ends of six axes in Table 4.

Table 4: Comparison of an ideal-typical characterisation of a command & control and an integrated & adaptive regime

Dimension	Command & Control Regime	Integrated & Adaptive Regime
Governance	Centralised, hierarchical, narrow stakeholder participation	Polycentric, horizontal, broad stakeholder participation
Sectoral Integration	Sectors separately analysed resulting in policy conflicts and emergent chronic problems	Cross-sectoral analysis identifies emergent problems and integrates policy implementation
Scale of Analysis and Operation	Transboundary problems emerge when river sub-basins are the exclusive scale of analysis and management	Transboundary issues addressed by multiple scales of analysis and management
Information Management	Understanding fragmented by gaps and lack of integration of information sources that are proprietary	Comprehensive understanding achieved by open, shared information sources that fill gaps and facilitate integration
Infrastructure	Massive, centralised infrastructure, single sources of design, power delivery	Appropriate scale, decentralised, diverse sources of design, power delivery
Finances and Risk	Financial resources concentrated in structural protection (sunk costs)	Financial resources diversified using a broad set of private and public financial instruments

The characteristics of integrated adaptive regimes have still to be regarded as working hypotheses since the change towards more adaptive regimes is yet slow and empirical data and practical experience thus limited. One possible reason for this lack of change is the strong interdependence of the factors stabilising current management regimes. One cannot, for example, move from top-down to participatory management practices without changing the whole approach to information and risk management. Therefore NeWater aims at a better understanding of the interdependence of key elements of water management regimes and the complex dynamics of transition processes in order to be able to compare and evaluate alternative management regimes and to implement and support transition processes if required.

The kind of changes needed is illustrated for the example of implementing the new water management paradigm "Living with water". Such a paradigm shift has been advocated in several countries (e.g. The Netherlands or Switzerland) as a consequence of the insight that current management practice are not sustainable in the light of increasing uncertainties caused by e.g. climate change. This implies a transition from the current flood control to a more integrated approach with multi-functional and dynamic landscapes as represented in Table 5. Resistance can be expected from actor groups who lose power. With new actors entering the scene the question of accountability for failed management notions needs to be addressed.

Table 5: Management regimes in regulated and controlled rivers compared with a future state with multi-functional and dynamic landscapes.

	Current state with regulated and controlled rivers	Potential future state with a multi-functional dynamic landscape
Stakeholder groups and their roles	Authorities as regulators in a highly regulated environment Engineers who construct and operate dams, reservoirs and levees Environmental protection groups fighting for floodplain restoration Insurance companies selling insurances against flood damage House owners living in floodplains Agriculture using land in the vicinity of rivers Shipping industry interested in well functioning water-ways	Authorities act as contributors to an adaptive management process with shared responsibilities Neutral third parties act as facilitators of the decision making process Landscape architects Engineers who have skills in systems design and cooperate with ecologists Environmental protection groups Insurance companies House owners with property in a floodplain at a higher risk of being flooded Tourism industry and tourists using the floodplains for recreation
Stakeholder participation	Little stakeholder participation – occasional consultation where different stakeholder groups and the public at large are asked to give their opinion on a management plan or scenario that has already been prepared by experts	Stakeholders and the public are actively involved in river basin management. This can be described as co-production of knowledge and co-decision making. Involvement can range from discussions with the authorities and experts, to actively contributing to policy development (co-designing), influencing decisions (co-decision-making), or even full responsibility for (parts of) river basin management.
Paradigm of water management	Management as control. Technology driven. Risk can be quantified and optimal strategies can be chosen. Zero-sum-games in closed decision space Implementation of controllable and predictable technical infrastructure (reservoirs, dams) based on fixed regulations for acceptable risk-thresholds	Adaptive and integrated water management. "Living with water". Acceptable decisions are negotiated. Implementation of a multi-functional landscape and increased adaptive capacity of the system. Designed risk dialogue and cascade of adaptation measures to live with extremes. Increased importance of real time forecasting systems
Institutional setting and governance	Institutional fragmentation Flood protection, nature conservation, regional planning and water management are often located in different authorities. Even the European Water Framework Directive does not address flood management. But it asks to preserve and/or restore the good ecological state of freshwater ecosystems. This includes the restoration of floodplains and will thus directly interfere with flood protection.	Polycentric governance and better institutional interplay Better horizontal and vertical integration of formal institutional settings to overcome fragmentation which might imply new institutions such as river basin management panels with defined responsibilities and decision making capabilities Stronger role of informal institutions and participatory approaches
Adaptive capacity	"Hard" approach to systems design, which has aims to implement long-lasting optimal solutions. Adaptive capacity is in general quite low due to high investment in infrastructure and often inflexible legal regulations (e.g. water use rights allocated for decades, technological norms that prescribe good practice and prevent innovation and change to new management practices)	"Soft" approach to systems design allows new insights to be taken into account, including responses to changing environmental and socio-economic boundary conditions. This is more in line with the new paradigm of adaptive water management.

The need to increase the ability of water management to respond to unexpected developments has received urgency with the increased awareness of the impacts of climate change. However, adaptive management seems to be limited in prevailing designs, practices and ideas surrounding river basin management. Reframing issues of river basin management to include climate change scenarios may aid in shifting the focus e.g. from flood management to a wider basin management view that includes storage and buffering of flow and capacity upstream. The gap between political rhetoric and operational management may be attributed to the neglect of the major changes needed to implement change. Social learning processes need to be actively supported (Pahl-Wostl *et al.* 2006).

It is yet too early to draw final conclusions from the current findings of the NeWater projects. However, it is already clear that the results will not deliver blue-prints for the "ideal" adaptive management regime but rather guidance on how to move from a participatory assessment of the current state to the design of a process for the development and implementation of appropriate management approaches.

2.4 CONCLUSIONS AND RECOMMENDATIONS FOR POLICY AND MANAGEMENT

The four projects described above bring forward a number of pertinent experiences and some evidence to inform the debate on new paradigms for water management. In reviewing these initiatives we have been necessarily selective in our reporting, focusing on those project outputs which help us understand the potential of models such as IWRM and AM. It is also true that the projects themselves were not undertaken with the sole purpose of evaluating new paradigms. Each project has generated multiple outputs and each has been influential within its own terms of reference. However, there are three particular areas where we feel that general conclusions can be made.

Uncertainties and their role in policy and management

Novel approaches to water management, which provide an alternative perspective on managing uncertainty, remain underexploited. The prevailing approach is still the scientific-technical paradigm for analysis combined with a control paradigm for management. This implies that uncertainties need to be reduced or at least quantified. There is little awareness of approaches that emerge from the social or management sciences where uncertainties may be perceived as opportunities and where the dependence on framing and the perspective on policy and management process is emphasised. Where new approaches such as AM and IWRM have been considered, there remain significant challenges in ensuring that the use of multiple management models is complementary and not in conflict.

Complexity and the dynamics of human-technology-environment systems

There is some evidence that points towards significant changes in management paradigms, which seek to encompass such multi-phenomena processes. A mechanistic worldview is slowly being replaced by a more organic, system-based perspective recognising dependencies, emergence and evolutionary dynamics. This has strong implications for the understanding of what management means.

Learning and managing change

There has been rapid development in actors' ability to identify barriers to change and develop participative planning and change processes. Insights also highlight that desirable change is predominantly hindered by existing governance patterns and that desirable futures may not always be in line with current power structures. Any approach to learning is obviously strongly dependent on the previous two aspects – how uncertainties are taken into account and dealt with and how well complexity and complex dynamics of human-technology-environment systems are understood.

These conclusions suggest that the following recommendations may be useful for management practice:

1. The complex socio-ecological nature of river basin environments and the inherent uncertainties associated with their management, in particular the impacts from climate change, have to be taken into account in policy development and implementation.
2. Selected management strategies should be robust and perform well under a range of possible, but initially uncertain, future developments.
3. An increase in, and maintenance of, the flexibility and adaptive capacity of water management regimes should be a primary management goal.
4. Effort has to be devoted to building and sustaining trust and social capital for problem solving and collaborative governance.
5. Entrenched perceptions and beliefs block innovation and change. Space has to be provided for creative and out-of-the-box thinking
6. Need to train a new generation of water management practitioners skilled in participatory system design and implementation.

The four research projects reviewed in this chapter provide some tantalising glimpses of the concepts and processes, which characterise new approaches to water management. But, in closing, we would note that our comment in the introduction to this paper that '*the gap between political rhetoric and operational implementation remains large*' is intended to flag up a weakness in research rather than politics. Widespread operational implementation of new management paradigms for the water sector will only come to pass when a robust case for intervention has been established. Such a case might include empirical and reasoning components and researchers need to be able to distinguish where evidence supports theory and where it questions (old or new) paradigms and models.

2.5 REFERENCES

Arentsen, M.J., Bressers, H.Th.A., and O'Toole, L.J. (2000) *Institutional and Policy Responses to Uncertainty in Environmental Policy: A Comparison of Dutch and U.S.Styles* Policy Studies Journal 28 (3), 597-611

Biswas, A.K. (2004) *Integrated Water Resources Management: A Reassment.* Water International 29 (2), June: 248-256

Bouwen, R. and Taillieu, T. (2004) '*Multi-Party Collaboration as Social Learning for Interdependence: Developing Relational Knowing for Sustainable Natural Resource Management'*.Journal of Community and Applied Social Psychology, 14:137-153.

Buttel, Frederick H. (1986) Sociology and the Environment: The Winding Road toward Human Ecology, International Social Science Journal 38: 337-356.

Collier, U. (1997) *Sustainability, Subsidiarity and Deregulation: New Directions in EU Environmental Policy*, Environmental Politics 6, 2: 1-23

Craps, M. (Ed.*) (2003) Social Learning in River Basin Management.* Report of workpackage 2 of the HarmoniCOP project (www.harmonicop.info)

Galaz, V. (2005) *Social-Ecological Resilience and Social Conflict – Institutions and Strategic Adaptation in Swedish Water management.* Ambio, 34(7). 567-572

Georgescu-Roegen, N. (1982) The crisis of resources : its nature and its unfolding. in Gregory A. Daneke, ed., Energy, Economics, and the Environment : Toward a Comprehensive Perspective, Lexington, Mass., Lexington Books, pp. 9-24

Global Water Partnership (2000), Towards Water Security: A Framework for Action. Global Water Partnership, Stockholm

Gunderson, L.H., and C.S. Holling. (eds). 2001. Panarchy: Understanding Transformation in Human and Natural Systems. Island Press, Washington, DC.

GWP-TAC (Global Water Partnership Technical Advisory Committee) (2000): I*ntegrated Water Resources Management.* TAC Background Papers No. 4. (GWP, Stockholm, Sweden)

Hilborn, R., C. J. Walters, and D. Ludwig (1995) *Sustainable exploitation of renewable resources.* Annual Review of Ecology and Systematics 26:45-67.

Holling, C.S. (Ed) (1978). Adaptive Environmental Assessment and Management. John Wiley & Sons: Chichester.

ICWE (1992) *Development Issues for the 21st Century. The Dublin Statement and Report of the Conference*, Dublin, Ireland, 26-31 January 1992. Geneva: World Meteorological Organization

Irwin, L. and Wigley, T. (1993) *Toward an experimental basis for protecting forest wildlife.* Ecological Applications 3(2): 213-217

Jeffrey, P. and Gearey, M. (2006) Integrated water resources management: lost on the road from ambition to realisation? Water Science & Technology 53 (1) pp 1–8

Jeffrey, P. and McIntosh, B. S. (2007) Description, diagnosis, prescription: a critique of the application of co-evolutionary models to natural resource management. Environmental Conservation. 33 (4) 281-293.

Jonker, L. (2002) Integrated water resources management: theory, practice, cases. Phys. and Chem. Earth. 27. 719-720.

Kaika, M. (2003) *The water framework directive: a new directive for a changing social, political and economic European frame-work.* European Planning Studies 11(3): 299–316

Kay, J.J. (1997) *Some notes on the ecosystemapproach: ecosystems as complex systems.* Pages69-98 in T. Rowley and G. Gallopin, editors.Integrated conceptual framework based on complexsystems theories. Working Document 167. CentroInternacional de Agricultura Tropical, Cali,Colombia

Kluge, T. (2005) *Kritische Betrachtung des Ansatzes „Integrated Water Resources Management"(IWRM).* In: Neuber, S., et al. (eds) (2005) Integriertes Wasserressourcen-Management (IWRM) - Ein Konzept in die Praxis überführen, p 31-44, Baden-Baden

Kraemer, R.A. (1998) *Subsidiarity and water policy.* In Selected Issues in Water Resources Management in Europe, Correia FN (ed.). Balkema: Rotterdam.

Lankford, B.A. and J. Cour (2005) '*From Integrated to Adaptive: A New Framework for Water Resources Management of River Basins' in the Proceedings of the East Africa River Basin Management Conference,* Morogoro, Tanzania, 7-9 March 2005

Lee, K.N. (1999) *Appraising adaptive management.* Conservation Ecology 3(2): 3. URL: http://www.consecol.org/vol3/iss2/art3

Lee, K.N. (1993) *Compass and gyroscope: integrating science and politics for the environment.* Island Press, Washington, D.C.

March, J.G. (1978) *Bounded rationality, ambiguity and the engineering of choice.* The Bell journal of economics, I 978g,No.2,587-608.

Mostert, E. (2003) *The European water framework directive and water management research,* Phys. Chem. Earth 28 (2003), pp. 523–527

Mostert, E., Pahl-Wostl, C., Rees, Y., Searle, B., Tàbara, D., and Tippett J. (2007) Social learning in European river-basin management: barriers and fostering mechanisms from 10 river basins. Ecology and Society 12(1): 19. [online] URL: http://www.ecologyandsociety.org/vol12/iss1/art19/

Newig, J., Pahl-Wostl, C. and Sigel, K. (2005) *The Role of Public Participation in Managing Uncertainty in the Implementation of the Water Framework Directive.* European Environment., 15, 333-343.

NRC (2004) *Managing the Columbia River: Instream Flows, Water Withdrawals, and Salmon Survival.* Washington, D.C.: The National Academies Press

Ohlson, D.W. (1999) *Exploring the application of adaptive management and decision analysisto integrated watershed management.* http://www.scarp.ubc.ca/thesis/ohlson/

Pahl-Wostl, C. (2002) *Towards Sustainability in the Water Sector - The Importance of Human Actors and Processes of Social Learning.* Aquatic Sciences: 64, 394-411.

Pahl-Wostl, C., Berkamp, G. and Cross, K. (2006) Adaptive management of upland rivers facing global change: General insights and specific considerations for the Rhone basin. Paper prepared for the International Rosenberg Forum on Water Policy, Banff, Canada, September 2006

Pahl-Wostl, C., (2007a) *The implications of complexity for integrated resources management.* Environmental Modelling and Software. 22, 561-569..

Pahl-Wostl, C., (2007b) *Transition towards adaptive management of water facing climate and global change*, Water Resources Management. 21(1), 49-62.

Pahl-Wostl, C., Sendzimir, H., Jeffrey, P., Aerts, J., Berkamp, G. and Cross. K. Managing Change towards Adaptive Water Management through Social Learning, Ecology and Society, in press.

Parma, A.M., Amarasekare, P. and Mangel, M. et al. (1998) *What can adaptive management do for our fish, forests, food and biodiversity*?. Integr. Biol. Issues News Rev. 1:16–26

Prato, A. (2003) *Adaptive Management of Large Rivers with Special Reference to the Missouri River.* Journal of the American Water Resources Association. Vol. 39 (4) pp.935-946.

Reefsgaard, J.C. et al. (2005). *Harmonised techniques and representative river basin date for assessment and use of uncertainty information in integrated water management (HarmoniRiB).* Environmental Science & Policy, 8, 267-277.

Robinson, J. B., Francis, G., Legge, R. and Lerner, S. (1990) Defining a Sustainable Society: Values, Principles and Definitions. Alternatives. 17. (2). pp36-46.

Selfa, T. (2004) Mobilizing at the frontier: Framing social justice claims in a natural resources management project in Rondônia, Brazil. Society and Natural Resources 17 (8), pp. 717-734

Smith, A. D. M. and Walters, C. J. (1981) *Adaptive management of stock recruitment systems.* Can. J. Fish. Aquatic Science 38 690-703

Tippet, J, Searle, B., Pahl-Wostl, C. and Rees, Y. (2005) Social Learning in Public Participation in River Basin Management. Environmental Science & Policy, 8(3), 287-299.

Walters, C. (1986) *Adaptive management of renewable resources.* McGraw Hill, New York.

Walters, C. J., and Green, R. (1997) *Valuation of experimental management options for ecological systems.* J Wildl Manage. 61:987–1006.

Walters, C.J. and Holling, C.S. (1990) *Large scale management experiments and learning by doing.* Ecology 71: 2060-2068

Winder, N. McIntosh, B. S. and Jeffrey, P. (2005) *The Origin, Diagnostic Attributes and Practical Application of Co-Evolutionary Theory.* Ecological Economics. **54** 347-361

3

Participation in water management: Theory and practice

Geoffrey D. Gooch and David Huitema

3.1 INTRODUCTION

Although shifts in water management in various countries and parts of the world may be very different in character, they do however seem to have one thing in common, which is, that the scope and ambitions of water management has increased. This implies that water management is now concerned with more issues, and affects more people than it did in the past (i.e. water management involves not just farmers but also industrial interests, nature conservationists and ordinary citizens). Water management has increasingly become concerned with *distribution*, not only of water but also of water related risks and benefits, and has therefore moved more and more into the realms of politics, where the traditional definition of what politics is can be formulated as 'the authoritative assignment of values' (or who gets what, when, and how). In the course of their profession, water managers have in fact always engaged in the assignment of values and their actions have therefore always been to a large extent political. However, whereas water management has traditionally been considered a predominantly technical field in many countries, the increase in scope, and new demands on water management, make the political character of the field very clear today, even to the most superficial observer and even in places where this might least be expected. These developments and the move in water management from the primarily technical to a combination of technical and political-social spheres has lead to an increased interest in public and stakeholder participation in water management. Many EU financed projects in water issues during the last years have mentioned participation, sometimes in connection with studies of water quality, of decision-support systems, and of water quantity. It is therefore of interest to analyse how this participation has been organised, who has been involved, and at which stage of the project process.

It seems clear that the relevance and saliency of public participation in water management in Europe has increased in the past decades. There are many reasons for this development. Some of these are on what one may call the 'demand side' of public participation: because there is more societal attention to environmental goods such as clean water and land- and waterscapes, the desire to participate on the part of the public has increased. Similarly, water management is increasingly touching upon politically sensitive issues such as land use planning, flood control, irrigation and even aesthetic uses of water. On the 'supply side', we may observe an increase in the number of institutional venues for public participation. In some European countries, such as the Netherlands, water management activities became subject to land use planning in the 1970's and authorities started experimenting with public participation. Many new interest groups also used the land use planning arena as an entry point to the world of water management. Increasingly faced with public opposition against water infrastructure projects in the 1980s and 1990s, water managers were interested in discovering public opinion at an early stage as they wanted to prevent opposition to their strategies and related court

© 2008 IWA Publishing. *The adaptiveness of IWRM; Analysing European IWRM research* Edited by Jos G. Timmerman, Claudia Pahl-Wostl and Jörn Möltgen. ISBN: 9781843391722. Published by IWA Publishing, London, UK.

cases later on. This interest in public opinion was not however simply based on a perceived means of defence; water managers also began to realise that they could learn from members of the public. The usefulness of contributions from the public to water management processes may vary, yet even without the benefits of a technical or natural science education, citizens often posses 'local knowledge' or 'ordinary knowledge' related to their being on their land that water managers need to incorporate in order to make accurate decisions. Faced with an increasingly complicated and uncertain environment, there is also a realisation of the need to learn to work together in order to successfully manage water issues (Ridder *et al.* 2005).

Yet while the issue of public participation in water management has come to be seen as more and more important, systematic studies of different types of participation, and of the outcomes of participation, seem to be few. There are also a limited number of examples of successful participatory processes in water management; the topic is referred to more often than it is implemented. In this chapter we therefore present an overview of the main issues that need to be addressed, as well as a discussion of the underlying ideas that should be taken into account when planning and implementing participatory methods in water management. First we present a discussion of different forms of participation and their role in representative and participatory democratic systems; we then continue with descriptions of a number of different participatory techniques. The chapter concludes with an analysis of participatory methods used in a small selection of EU financed research projects focussing on water management. Our ambition has not been to conduct a comprehensive review of EU financed projects, but to illuminate some of the problems facing public participation by providing examples of attempts to manage these problems.

As our focus is on participation, we also touch upon the role of institutions in water management. Our overview of EU projects conducted for this chapter (see below) has shown that institutions, both formal and informal, are important. Institutions, in the form of organisational contexts and/or the values and norms of water managers and the public, play important roles in the ways in which the public can, and are prepared to, participate. An immediate question that springs to mind here is which of these are most important for the success of participation in water management? Are the norms and values of the stakeholders and public foremost, or are the formal structures that they are allowed to participate within most important? This is far from merely a simple question, as the choices of the participants, as well as the institutional contexts that they work within, create a wide range of alternatives that water managers must take into account. As will be shown later in the chapter, our selection of potential participants is not always completely in our own hands; self-selection processes and willingness to participate also influence our alternatives. Participants may thus well come to the participatory process with ideas and values that take us by surprise. In a similar way, the formal institutional contexts, the organisational form for the participation, can also strongly influence the outcomes of participatory processes.

The distinction between formal and informal institutions is connected to the continuous debate in the social sciences as to whether explanatory factors for human action can be found foremost in structures, or in actors. Different theories provide, as we would expect, different answers to this question and we will not go in depth into the debate here. Bas Arts (Arts 2000) attempts to bring the two main different views (actor or structure) together in a way similar to that of Anthony Giddens structuration theory (Giddens 1991). Arts' model takes the issue at stake as the central point of departure, analyses how rules are defined around the issue, and by which agents, how these are structured and the extent to which these rules shape the conduct of agents. This model therefore takes into account the mutual interaction of both structures and actors and Arts suggests a number of steps when using his model, which are:

(1) Definition of the issue concerned (what's the problem?)
(2) The identification of relevant principles, norms, rules and procedures that define the issue
(3) The identification of the relevant agents
(4) The reconstruction of relevant routines, intentions and capabilities of these agents
(5) The assessment of the distribution of capabilities among them (Arts 2000).

This relatively straightforward model can also be of help when formulating participatory strategies for water management. Participation for participation's own sake may appeal to democratic theorists, but in the often pragmatic and resource-limited world of water management there is usually a need to take a more concrete problem as the point of departure. What is the issue at stake? Is it water supply to urban areas, the quality of drinking water, problems with waste water treatment, or something else? The problem at hand will determine the degree and type of participation. The second step leads us to ask the question of the form and outcome of the participation. Do the rules (political and/or legal) allow the participants advisory functions only, or can the participatory process lead to decisions, to changes of policy directly influenced by the participants? As we stress later on in this chapter, it is very important to make this clear for everyone involved right from the outset. The third step is to determine who the relevant agents or actors are. Who should be included in the participatory process, and why? While there are usually a number of obvious stakeholders involved in a specific issue, it can also be necessary to think 'outside of the box', and to be prepared to include other, less traditional actors in the participatory process. The final steps of the model involve the re-reformulation of routines in order to allow the public and stakeholders to participate in a realistic manner, to enable water managers to utilise new insights, and to consider how the competences of those who have participated can be utilised in management and the implementation of policies. This is often the most difficult, as our discussion on institutions will show. The identification of the issues, norms and rules, agents etc. can be used as an input into the determination of the most suitable method for participation, and be used as a point of departure for the discussion presented in more detail in Section 2.3 of this chapter.

Participation is of course only one of the challenges for water management, problems arising from the production of scientific information, the communication of that information, legal systems, and institutions are also important (Gooch and Stålnacke 2006). The question of information, and its communication, is also central for public participation, at least when we talk about *informed participation*. As we have observed, rapid changes in the driving forces behind water problems have also increased the levels of uncertainty under which policies must be formulated, and have increased the (always present) visibility of political forces and considerations. While this in turn has lead to an increased awareness of the necessity of including stakeholder and public participation in water management decision-making, as well as in other areas (OECD 2001), it is also important to stress that reliable information must be made available to the public in an understandable form. While we note that local knowledge can also contribute to water management processes, it is vital that public participation is also based on an understanding of the knowledge provided by the scientific, engineering, and management communities. A method which is based on the engagement of experts and the use of scientific knowledge by the public is described later in this chapter in the section on citizen juries.

Another aspect of participation that needs to be taken into account is the level (administrative and geographical) on which the participation takes place. Participation at the local level is often easier to organise as the public often have both knowledge and interest in local water management issues. The higher and larger the spatial and administrative level, the more difficult it may be to provide the necessary information and to motivate participation. One way of coming round this problem that has been attempted in the United States is to organise local participatory processes and then to bring representatives of these local processes together. Spatial levels are also important for the role played by administrations. Government at all levels was seen as an important actor in an earlier overview of water management literature (Höglund and Gooch 2002) and our overview of EU research projects in this chapter has shown that local and regional government are still important actors in water management. Regional government seems to be especially important, probably because in many countries it is this level of administration that manages water and environmental issues. The EU Water Framework Directive also takes a regional perspective, albeit based on hydrological and not administrative regions. National governments also play, of course, an important role in water policy, especially in Transboundary cooperation such as on the Rhine (Dieperink 1998), yet many other actors now play important roles even in such conditions (Nachtnebel 2000). At the state level national authorities are now dependent on the compliance of actors they cannot force by traditional governmental measures. Networks have come to play a crucial role in governance process, and networks also play an important role in the dissemination of knowledge. Within these networks, civil society, stakeholders, and even the public, have come to play important roles. How can we explain this? First, from the traditional three levels of government and administration, national, regional and local, the development of the EU has added a fourth level of governance. The importance of the different levels varies among the European countries, but the influence of the EU has increased during the last decades of the 1900's, and the EU has encouraged developments at the regional level (Gooch 2000), although not always in ways that we would expect. Pierre and Peters argue that states' possibilities to govern have changed since they've become more interdependent, partly due to the process of globalisation which has strengthened the global level of governance, and international organisations like WTO and OECD are growing in importance (Pierre and Peters 2000). In the field of water management there are national laws, EU directives, the Helsinki convention and other international conventions. The implementation of many of these laws has to be conducted on the regional and local levels, and it is on these regional and local levels that the practical work can be done. Aynsley Kellow puts it this way: 'International policy must first be given effect by domestic legislation, and then that policy must be implemented' (Kellow 1997). In an analysis of water management we might consider five levels of decision-making/implementation, the global level, the EU level, the national level, the regional level, and the local level. Implementation occurs at the regional and local levels, and this can probably explain the perceived importance of these levels of government in the project survey. The existence of these levels constitutes a major challenge for public participation, as the possibility of effective participation, as well as the methods used, differs at the different levels.

We have pointed to some of the arguments for participation. These include that way that participation is expected to lead to greater legitimacy for the policies (due to the spread of responsibility), and to improvements in policy implementation (due to a readier acceptance of policies in which stakeholders and the public have participated). Citizen participation is also important for democratic decision-making at large (Dryzek and List 2003), as a possible way to counter the fact that large sectors of the public *'have grown distrustful of politicians, sceptical about democratic institutions, and disillusioned about how the democratic process functions'* (Dalton 2004). As we have noted, a major question posed by participation is who should take part in the participatory/deliberative process? Must we all gather together to decide water management issues? Hardly; according to Gastil (Gastil 2000) only a minority of democratic theorists argue that deliberation needs to take place in a face-to-face setting among members of the public. The costs of deliberative processes and the potential development of e-democracy (Gastil 2000) point perhaps in other directions. However, deliberative processes such as citizen juries can provide policy-makers with opinions and input from a group of well-informed members of the public (Crosby 1998), and the problems associated with the method need to be addressed practically as well as theoretically.

3.2 CITIZENS, CIVIL SOCIETY AND PUBLIC PARTICIPATION IN WATER MANAGEMENT

Changes in water management paradigms, such as the WFD, necessitate changes in decision-making procedures, and the development of new forms of governance in water management. The involvement of citizens and civil society has come to be seen as necessitating not just information and consultation but a form of *democratic* governance, involving the participation of civil society in political and policy processes. However, while this implies that the involvement of the public and the development of civic engagement are central to governance (Jun 2002; Smith and Ingram 2002), the idea of civil society is often used more as a normative than an empirical concept (McLaverty 2002). Exactly what it involves to be a member of civil society is somewhat unclear and this lack of clarity creates problems when it comes to deciding on how 'civil society' should be included in participatory processes. Hall claims that 'civil society *is* complicated, most notably in being at one and the same time a social value and a set of social institutions' (Hall 1995). Gellner defines civil society as a cluster of institutions and associations, entered into freely, that can prevent excessive use of state power (Gellner 1995), while Cohen and Arato describe civil society as being between the state and economic interests (Cohen and Arato 1994). For our purposes civil society will be defined as a sphere of activity separate from the politico-administrative and business sectors, a sphere that includes certain NGO's as well as members of the public in general.

3.2.1 The concept of public participation and its interpretations

Many studies of water management stress the state as the main actor. While we agree that the state is important, we also claim that non-state actors and individuals have increasingly become more central as they represent interests, focus attention on specific issues and induce state actors to work towards the creation of new institutional arrangements. To observe other actors than states is thus crucial if we are to understand the nature of the dynamics of water governance, and the debate on governance leads us to understand the important role of civil society (Gooch 2006), and that:

At the global level, governance has been viewed primarily as an intergovernmental relationship, but it must now be understood as also involving non-governmental organisations (NGOs), citizens' movements, multi-national corporations, and the global capital market (Governance 1995).

The participation of NGO's and citizen movements is now seen as an integrated aspect of governance; however, as public participation is often defined in radically different ways we first need to clarify our understanding of the meaning of the term before going on to analyse its practical applications. Here we define public participation as the taking part, by ordinary citizens or their collectives, in the processes of government and/or governance; we refer to situations in which a substantial number of citizens play a part in the process by which leaders are chosen and policies are shaped and implemented (Birch 1993). What exactly we mean by 'taking part' and 'playing a part' will be discussed later, as it can involve a number of different forms of participation. Regardless of the form that the 'taking part' takes, public participation is increasingly regarded as an important element of a well functioning democracy. The fact that citizens in a certain country or region can vote every so often (usually every four to five years) is now often merely seen as a minimum requirement for a government system to be called democratic. While official attitudes to public participation have differed during the last decades, public participation is now entrenched in discourse and law, and the absence of participation in decision-making processes is usually considered illegitimate (Bulkeley and Mol 2003). Direct public participation is however not the only type of 'taking part'. Other forms of participation include membership of a political party, membership of a pressure group or organisation, community action, civil disobedience, canvassing in elections, etc. In this chapter we are however specifically interested in direct citizen participation in the policy-making processes of water management, and this is the topic that we focus upon.

Within the water sector, an important advance in European water management in recent years is the EU Water Framework Directive. The Water Framework Directive (WFD), aims at guaranteeing good water quality and sufficient water quantity in the European Union (EU), and combines a number of earlier legislative acts into a single directive, which must, according to EU law, be implemented in all the member states by 2015. It is based on hydrological and not existing administrative boundaries, which, apart from in a few countries such as France, itself is quite an innovation in water management. It also represents the first attempt to create a European (EU) wide harmonisation of water management practices. While allowing individual countries flexibility in exactly *how* the aims are to be achieved, the WFD implies a greater role for the European Union in water management, specifically through a set of coordinated planning processes at the level of certain large river basins. The framework directive also creates the need of all interested parties to be involved in the production and also in the review and updating of the river basin management plans. Through the hydrological units that must be formed, the WFD also necessitates the creation of new administrative units. The WFD is important in the context of this chapter as public and stakeholder participation are specifically stipulated in the directive.

As regards the implementation of the framework directive and the proposed system of water management the WFD also stipulates not only information but also the active involvement of the public and stakeholders and the development of methods to do so. The preamble to the WFD acknowledges that: '*the success of the Directive relies on close cooperation and coherent action at Community, Member State and local level as well as on information, consultation and involvement of the public, including users*' (EU 2000). From this quotation it becomes clear that the WFD distinguishes between three levels of participation, with an increasing degree of control over the outcomes of the process: these are 'information', 'consultation', and 'active involvement'. Information can be provided by means of brochures, briefings, press releases, etc. Here there is a one-way flow of information, downwards, from the initiators to the stakeholders. The stakeholders and public (or publics) do not themselves have the opportunity to give their opinion, as they only 'consume' the information that is offered to them. 'Consultation' can be considered a low level of participation. Examples of 'consultation' are interviews, opinion polls, and public hearings. The flow of information is still one-way, but this time upward, from the stakeholders and the public to the initiator. In both 'information' and 'consultation', there is no, or only limited, interaction within the stakeholder group, and between the water managers and the stakeholders. 'Active involvement', on the other hand, can be considered a higher degree of participation. 'Active involvement' can be organised by means of citizens' juries, planning cells, consensus conferences etc. We will discuss and develop these different forms of participation in the following sections.

3.2.2 Public participation goals and design, methods, and evaluation criteria

We will now move on to discuss the reasons for participation and a typology of participatory techniques, in other words, how the public can 'take part' in water management processes. Public participation is an 'essentially contested' concept, which means that its meaning will always be debated. This is partly because public participation can be approached from many different perspectives. In this chapter, we adopt a functional perspective, that is, we assess whether public participation achieves certain preset goals and we do not plead for increased participation for the sake of participation itself (Coenen *et al.* 1998).

Elsewhere (Van de Kerkhof and Huitema 2004) the point has been made that all arguments for public participation in environmental management essentially relate to five goals. First, public authorities may want to use participatory methods for *raising awareness of environmental issues*. While the environment as such has kept its focus in the public mind, individual environmental issues tend to follow the 'issue attention cycle' (Downs 1972) which implies that no matter how serious a certain issue, public attention for it will wane and public authorities need to do their part to keep issues in the public eye.

A second goal of participation is to reach *better decisions* by including information about facts and consequences of the intended measures into the decision-making process and by allowing the environmental manager to find feasible (in the sense of acceptable and workable) solutions. Certainly if managers succeed in creating a 'safe atmosphere' and cooperation with the participants, these may bring forward new information that can be beneficial to the decision process. Participatory exercises can be expected to show what, according to the citizens, should be the priorities of environmental management, and what kind of choices should be made. These may be important insight for managers. Citizens may ask questions, sometimes very fundamental ones, about understandings and agreements that all insiders know about but which may look strange from the outside. The least that one can take from this is cues for better communication with the public.

The third possible goal of participation is to increase the *legitimacy of decision making,* by reducing the distance between environmental managers and the public, by engaging the public in deliberation about the decisions that need to be taken and by generating support for measures to be taken. Participation can provide people with a feeling of control. Policies and measures that are produced with people's cooperation and for which people feel a level of responsibility will increase their acceptance. In addition, it is a well-known phenomenon that involvement breeds further involvement, thereby reducing the perceived distance to decision makers.

A fourth goal of public participation is to increase the *accountability of decision making*, which is about the government becoming more responsive to the concerns of the public, about creating transparency and giving access. Government actions and decisions need to be motivated, preferably by referring to arguments that strike a chord with the public and address potential issues and concerns they may have. Public participation is one way to gain access to the public mind and find out what it is that people care about; this is the responsiveness aspect. At the same time, public participation can fulfil a function in showing the public what the bureaucracy and politicians are doing and why they are doing it. Preconceptions about political and bureaucratic behaviour are aplenty and not always correct, and public participation can help to solve certain misconceptions in this field (transparency). Open access of the people to public institutions is needed to give them a share of ownership ('this is my public authority') and create a sense of trustworthiness ('they have nothing to hide').

Finally, participation may result in *collective learning*. Talking to stakeholders and citizens will create awareness and understanding of different perspectives on the issues at hand for environmental managers. In addition, stakeholders, government and scientific experts ideally enter into a dialogue and, by interaction and debate, learn how to collectively manage a natural resource (Pahl-Wostl 2002) and get insight into conflicting views and interests. If this succeeds, one could say that a collective vision has been created, one that can guide societal action from then on. Obviously the scope of such a learning process may differ; perhaps it is more likely that collective learning will occur within the group of participants or between participants and experts rather than collective learning occurring in the entire community.

3.2.3 The design of public participation exercises

Because of the importance of the relationship between participation and representative democracy, the degree to which the outcomes of participatory processes are binding has attracted attention from both academics and practitioners. Many discussions of public participation refer to Arnstein (Arnstein 1969) an author who was dissatisfied with the then existing forms of participation and proposed a normative 'ladder of participation'; this essentially suggests that public participation is only 'real' when the participants make binding decisions. Positioning public participation so directly against representative democracy (or the bureaucracy for that matter) is unlikely to increase the enthusiasm for public participation. It is also not *realistic* to focus on power alone, as there are other design parameters that are equally relevant. Imagine a participatory process with full-fledged authority to decide on a range of water management issues, but with no information whatsoever? Even though the participants are formally very powerful, lacking any information, they would have to refrain from taking decisions.

Instead of the traditional distinction provided by Arnstein, we propose that the water manager should take into account the following when planning participatory processes:

(1) Which competences should the participants have? Is what they are suggesting binding or not? Can the participants determine themselves which issues they will address or not? To which extent do they have to take into account and are they bound by existing policy choices. This is the classic focus of participation studies: what is the decision making power of participants? The greater this power, the greater the participative nature of the decision process. Here it is very important not to create unrealistic expectations; if the participation can only result in advice that might or might not be taken, it should be made quite clear to the participants from the start.

(2) Who should participate? What are the conditions of access to the process: who is in and who is out? It matters a great deal whether participation rights are allotted only to landowners, to tax payers, to organised interests, or to anyone having or showing an interest. The more liberal the rights of access, the more participatory the process becomes.

(3) What kind of information will the participants have access to? The more open access is to information, the more participatory the process will probably be. It is also a matter of defining the type of arguments that count and the ones that do not count. The stronger the emphasis on arguments related to technical and cost effectiveness issues, the less participatory the process becomes. This is because the type of arguments brought forward by participants tends to be either related to local knowledge or to normative evaluations. It is also important to determine how the information will be presented, and by whom.

(4) How much can it all cost? Do participants have to pay their own way or not? Costs associated with participation relate to the giving up of free time, the gathering of support and organising a collective, travel and sometimes the hiring of experts to help formulate and defend one's position.

Obviously, the choices that are made for each aspect of the design of a participatory approach can, and probably will, be related. Unless sufficient time and resources are allocated to provide understandable information to the public an emphasis on technical and economic factors may effectively rule out many potential participants, whereas covering the costs of participants may provide a positive effect on people and groups who would otherwise have difficulty taking time off work, etc. Fortunately, there is a set of 'pre-fabricated' methods available that imply choices on each of these parameters and we discuss these methods in the following section.

3.2.4 Public participation methods

Participatory methods can be understood as a group of procedures designed to involve, consult, and inform the public in order to allow them to provide an input into public decision making. The literature on participation is characterised by an overwhelming variety of methods and techniques. The methods range from those that canvass opinions to those that elicit judgments and decisions from which actual policy might be derived. In order to give the reader an idea of the 'menu' of participatory methods available for the practice of water management, Table 1 provides an overview of the main alternatives. Here we present a selection of methods that distinctively differ from one another with regard to the four aspects mentioned earlier: authorities assigned to the participants, the choice of participants, the flows of information; and the costs to participate in the process. The methods include referendums, public hearings, opinion surveys, negotiated rule making, web panels, citizens' juries, focus groups and social learning groups.

The table clearly demonstrates the differences between the methods. With regard to conditions of access ('who is in and who is out?'), a number of methods only allow for a small selected group of participants (citizens' jury, negotiated rule making, focus group, social learning groups), whereas other methods allow for a potentially large number of participants (referendum, opinion survey, web panel). In particular the referendum and opinion survey do not facilitate multiple flows of information between the participants, whereas most of the other methods do. Some of the methods, such as the citizens' jury and the social learning group, explicitly focus on facilitating multiple flows of information between the public / the stakeholders and (scientific) experts. In a focus group the direct flow of information from outside into the focus group during the process is rather low, although discussions often build on information obtained earlier from different sources such as the media. The reason for this is that the focus group aims to elicit the viewpoints and opinions of the participants in that particular group. The process should, therefore, not be influenced by information 'from outside' during the actual group discussions. Concerning the costs, in most of the methods the organisers (or their financers) bear the expenses to be made for organising the participatory exercise. In many cases the participants will have to pay for their own expenses. In some methods, e.g. the citizens' jury, the participants may receive a compensation for giving up free time. For a number of the methods, such as referendums and opinion surveys, there are hardly any expenses involved for the public. Also with regard to the authorities that are given to the participants the methods are rather distinctive. In a referendum and opinion survey, the participants do not have the authority to determine themselves which issues they want to address; these are predetermined by the organisers. In the other methods, the organisers usually define a theme or a question (a 'charge' in the case of a citizens' jury) within which the participants are free to choose which topics they want to discuss. With regard to the authority to make actual decisions, in the case of almost all the methods the outcomes are not binding for policy but serve as an advice to a public authority. A notable exception is the referendum, which can be binding for policy, and there are a few cases in the United States in which the outcomes of a negotiated-rule-making process were binding.

3.3 EXAMPLES OF PARTICIPATORY METHODS FROM EU-FINANCED RESEARCH PROJECTS

Following the sections above which laid out a number of principals and concepts that should be taken into account when planning participatory processes, we will now continue with three examples of research projects which have included a strong focus on participation.

3.3.1 MANTRA –East

The MANTRA-East project ran from 2001-2004, and studied transboundary water management on the Estonian-Russian border, around Lake Peipsi. It had three main objectives. The first of these was to evaluate the applicability of the EU Water Framework Directive in the future EU border regions. This included an assessment of the state of eutrophication (e.g. ecological status) in lakes and river basins, as well as the development of strategic lake and river basin tools for source apportionment, retention, and time-trends in nutrient loads. The second objective was to develop methods to improve communication and utilisation of scientific information. The third objective was to develop institutional mechanisms and policy instruments for decision making under conditions of transition and uncertainty, including institutional arrangements for public and stakeholder participation. The first objective was primarily the domain of the natural scientists in the project, the second objective was mostly managed by geographers and information specialists, and the third objective was the responsibility of the social scientists and policy analysts. The underlying rationale behind this model was that scientific information produced by the natural scientists was needed in the policy process, as well as knowledge of the policy process (decision-making and implementation) itself. This information needed to be communicated to the actors in the policy process, as well as to stakeholders and end-users. The project also needed input from the people affected, in other words, public participation.

Project methods

The natural science approach was based on quantitative methods and results that could be analysed through statistical methods and mathematic models. The communication of scientific knowledge was studied through a combination of information cycle methodology and qualitative analyses of communicative processes. The policy process, the involvement of stakeholders, and public participation were analysed using a heuristic approach in which qualitative interviews, surveys, focus groups, text analysis and institutional analysis were combined. The experience of the team was that understanding of each others viewpoints could, and was, increased by continuous discussions of the nature of the problems and the best ways to solve them. In this way the project adopted a problem-based methodology, attempting to elucidate the central aspects of water management from various viewpoints. The difficulties with the approach were, after an initial period, reduced and resulted in an increased understanding of differing scientific beliefs. The next step was to develop an approach that was synergetic, and which resulted in a greater understanding of the problems of water management than if each discipline had simply presented individual results which were then combined in a final report. The approach chosen was that of qualitative-quantitative-qualitative scenarios.

The MANTRA-East project needed to develop a methodology that could present the results of relatively complicated scientific modelling exercises to stakeholders and the public in a manner that they would be able to understand. Scenarios, through their simplification of this information, can make this process easier. Basically scenarios can be qualitative or quantitative, explorative or normative. Scenarios are not precise predictions of the future (Porter 1985), they should be seen as simulation tools, as a technique similar to, but different from, models (Jouvenel 2000) as scenarios attempt to construct general representations of possible alternative futures. In the MANTRA-East project, qualitative scenarios, 'storylines' or 'scripts', were first developed (Gooch 2003); these were narrative texts that were representations of possible futures. They did not contain detailed numerical information, and could be written in an easily understandable manner for policy-makers, stakeholders, and the public. Utilising past and present trends, the qualitative scenarios attempted to see which factors might lead to likely futures. The qualitative/explorative scenarios were then used as input into computer-generated models that can best be described as quantitative/explorative in the first stage, and quantitative/normative in a second stage. According to the project methodology, the results of the quantitative models were then used to modify the qualitative scenarios, and to develop qualitative/normative scenarios that provided grounds for policy recommendations. The time period chosen for the scenarios was twelve years, from 2003 to 2015.

Competences of participants

Referring to our earlier discussion of the four aspects of participation, competences, participants, information and costs, we will attempt to answer each question in turn. The answer to the first question, that of the competences of the participants, is that the degree of influence differed as the project included inputs from widely differing groups. Workshops were arranged for policy-makers and environmental managers at which they acted out role games and analysed possible future developments (Roll and Timmerman 2006), and these policy makers obviously had strong possibilities to influence water management in the region. On the other hand focus groups, interviews and other studies were conducted with members of the public, and here the outcomes could only be used in an advisory function. An important actor was the transboundary commissions, consisting of representatives from both Estonia and Russia, and the commissions played an important role in policy recommendations.

Identification of participants

Moving on to the second aspect of participation, the identification of the participants, the wide range has already been inferred in the previous section. The main stakeholders for water use and protection in the Lake Peipsi Basin were structured according to their geographic levels of operation. At the international basin level stakeholder groups included international organisations that were involved in development and implementation of policies for management of transboundary waters in Europe such as the World Bank, UNECE, UNDP, UNEP, Council of Europe, IWAC, WWF, the European Commission and other European Union organisations and agencies. In the Baltic Sea region, important international organisations included the HELCOM, Council of Baltic Sea States, and Baltic Sea Sub-regional Cooperation Council among others. On the true Lake Peipsi level there were two intergovernmental commissions, the Estonian – Russian Intergovernmental Transboundary Water Commission and the Estonian-Russian Joint Commission for the Use of Fish Resources of Lake Peipsi, Lake Lämmi and Lake Pihkva.

Table 1: A selection of methods for public participation

Method name	Authorities / competences	Conditions of access	Information flows	Costs to participate
Referendum	Voting mechanism, all participants have equal influence, outcome often binding	Potentially all members of a population, realistically a significant proportion	Information supply through the media, the method does not allow for interaction and debate	Costs borne by organiser, in general no, or very low, costs for the participants, except for travel expenses and the like
Public hearing	Topic is often pre-determined, outcome not binding	Interested citizens, limited number because of size venue	Presentation of plans in open forums, public may voice opinions, low level of interaction	Costs borne by organiser, in general no or very low costs for participants, such as travel expenses, giving up free time and the like
Opinion survey	Topics are predetermined, outcomes of survey not binding	Large samples, usually representative	Written questionnaire or telephone survey, used for information gathering, one-way flow of information and opinions, no interaction between participants	Costs borne by organiser, in general no, or very low, costs for participants
Negotiated rule making	General topic predetermined, information supply on participants' request, independent facilitator moderates discussion, outcomes most of the time not binding but serve as an advise to policy makers	Small number of representatives of stakeholder groups	Working committee of stakeholder representatives, high degree of interaction and flows of information between stakeholder groups, consensus required on a certain matter	Costs borne by organiser, in general no or very low costs for participants, such as travel expenses, giving up free time and the like
Web panel	General topic predetermined, participants may determine sub topics, outcomes not binding	Open forum, potentially all members of a population, realistically a significant proportion of interested citizens	Public may voice opinions and discuss electronically, flow of information to the participants often rather low	Costs borne by organiser, in general no or very low costs for the participants
Citizens' jury	General topic predetermined, information supply on participants' request, independent facilitator moderates discussion, outcomes most of the time not binding but serve as an advise to policy makers	12 to 24 members of the public, roughly to be a 'microcosm' of the population	High degree of interaction and flows of information between citizens and experts, consensus not required	Costs often born by organiser, but sometimes participants have to give up free time, and travel, without compensation
Focus group	General topic predetermined, participants decide upon sub topics they want to discuss, little input from facilitator, outcomes not binding	Small groups of 6 to 10 people, usually from similar backgrounds	Free discussion on general topic, which is recorded, used to asses opinions and attitudes, high degree of interaction between participants, flow of information often rather low	Costs often born by organiser, but sometimes participants have to give up free time, and travel, without compensation
Social learning group	It is up to the participants to set up experiments, measuring results, and conducting analysis, they are supported by scientists, outcomes not binding but serve as an advise to policy makers	Small number of representatives of stakeholder groups, accompanied by researchers	High degree of interaction and flows of information between stakeholders and scientific experts, consensus not required	Costs borne by organiser, in general no or very low costs for participants, such as travel expenses, giving up free time and the like

On the national level stakeholder groups included national governmental institutions (sectored national agencies, such as ministries of the environment, economics, etc.); businesses and farmers' associations; national NGOs and similar associations; national level universities, including researchers as well as members of the public and certain ethnic minorities. The experience of the project was that informing and involving stakeholders at a local level is important and can be most effective when concerning local environmental problems, such as water pollution, due to the local nature of such issues (see discussion above). One of the most important local level stakeholder groups was the local government which was responsible for water management within its territory, along with other local authorities, stakeholders such as schools, local businesses, women and youth groups, and grassroots NGOs. All were found to play important roles in the implementation of water management policies.

Information needs

The third step involved the identification of the type of information necessary and suitable for the problem at hand, which was as stated earlier eutrophication in Lake Peipsi. To assess the information needs of stakeholders several approaches were applied (Roll and Timmerman 2006). The needs of the international stakeholders were quite well-known through earlier activities. Thus, it was decided to place emphasis on the needs of the local level stakeholders through a small series of dedicated information needs assessment activities. The information needs assessment activities included group interviews in several municipalities with local stakeholders, interviews with local administrators in Estonia, an anthropological study in four rural municipalities (two in Estonia and two in Russia) aimed at understanding the public's perceptions of the environment and the willingness of local stakeholders to participate in discussions and decision-making processes with regards to environmental issues on the local level The process also included a survey of 200 – 300 people that explored the opinions of local people, and finally based on these public opinion surveys, 8 focus group discussions in local municipalities and 2 citizens' juries organised on the Estonian and Russian sides of the region. Furthermore, interviews with water management experts in eight regional environmental protection departments of Estonian Ministry of the Environment were conducted. Local communities and stakeholders address local water related environmental problems. Therefore the perceptions of these stakeholders concerning regional development perspectives and the environmental state were important to understand. Interviews with local stakeholders showed that communication on the local level often takes place through local newspapers of which some are distributed free of charge. Local meetings were considered important to develop a dialogue between local authorities and inhabitants. A few people used Internet, mostly those involved in work of local authorities and younger people. Access to Internet is higher on Estonian side. Many people stress problems with the physical communication infrastructure. Yet, in many municipalities there were already public Internet centres or a computer class with a permanent Internet connection, or these are planned to open.

In order to assess information needs and to provide information, a number of techniques were used in the project. These included Community Information Sessions and Campaigns, an interactive approach that not only worked to inform the general public on environmental issues, but that also provided a medium for suggestions, comments and participation. On a national and regional level, especially, effective measures for disseminating information to the public and integrating stakeholders in the policy making process included using information centres, public awareness and educational programs and cultural events. Another technique was the production and distribution of written information including brochures, information packets, pamphlets, leaflets and information booklets. Citizen Juries were also used (see below). In these interactive "town hall" type meetings, a randomly selected and demographically representative panel of citizens discuss the issues surrounding a particular policy area. It is of especial importance to the environmental field seeing how it is a way to collaborate both expert knowledge and public concerns, which usually is a challenge. Finally, the inclusion of an NGO in the project facilitated a variety of other methods, and the Peipsi Centre for Transboundary Cooperation coordinated various other activities and projects in the interest of furthering the public participation through the dissemination of information. These included fieldtrips, painting contests of water-related issues, conferences, workshops and tours.

MANTRA-East results

The MANTRA-East project attempted to combine a number of participatory methods (focus groups, surveys, citizen juries etc.), and to examine ways of presenting scientific information to stakeholders and the public. The use of maps and scenarios were ways through which physical-chemical and biological status and different possible futures were presented in a relatively easily understandable manner, and they were also used in discussions about future trends. In the case of the problem at hand, eutrophication, information played a central role, and it was vital that this information could be readily comprehended by stakeholders and the public.

3.3.2 River Dialogue

The River Dialogue project ran from 2003-2004, and focussed specifically on the use of participatory methods in the implementation of the EU Water Framework Directive. The project aimed to identify the best approaches to increasing public empowerment and involving the public in the implementation of the EU Water Framework Directive and river basin management plans through focus groups and citizen juries conducted in three European river basins: the Motala Stroem in Sweden, IJsselmeer in the Netherlands and the Emajõgi River in Estonia. It aimed at developing best practices for increasing public participation through the refinement and organisation of these methods in the three countries participating in the project, and to develop guidelines for the organisation of citizens' juries and other forms of participation. A central ambition was to promote a two-way dialogue on the issue of public consultations and involvement in river basin management, and to help improve the communication of scientific information from scientists to water management practitioners and the public. Behind this ambition was an awareness of the need to increase experts' and scientists' ability to learn local knowledge from the public. In the two following sections the methods used in the project, focus groups and citizen juries will be described.

Focus groups in River Dialogue

The first method used in the project was focus groups; these were organised in all three case study river basins and in all 25 focus groups were conducted with a total of 175 people participating in the groups. The organisation of the groups followed a common form, although some variation was allowed. Generally, the participants were recruited by first contacting a representative for one of the groups that the project partners were interested in. These included farmers, fishermen, students, members of a nature protection organisation, water managers etc. These contact persons then recruited between 6 and 10 participants for the focus groups. The interviews were recorded and transcribed, and then analysed.

Focus groups can be used both as the main tool in a research project, and in combination with other methods as in the River Dialogue project. For example, before a larger survey is sent out focus groups can be used to examine if the questions in the survey are constructed and understood in a way that the researchers would like them to be understood. Focus groups can also provide indications that help the researcher to develop the survey, for instance by giving insight into important areas that should be covered by the survey, but that the researchers have overlooked. Using a similar approach, researchers can use focus groups to validate the conclusions that are drawn from survey material. Focus groups can also be used before the researcher has concluded his own analyses of the material, the thought here being that the researcher can both control his own findings from the survey, and also obtain indications as to what kind of trends he or she should look for in the material. Focus groups can also provide a method suitable for getting a brief understanding of an area earlier not covered. By conducting focus groups within a possible field of interest the researcher can gain insights which may help to generate ideas on how to conduct continued research into the field (Wibeck 2000). Focus groups can also be combined with, for example, participant observations, individual interviews and text analyses. In connection with case studies which use a number of different techniques, focus groups can function as one of the preferred methods.

A focus group can be structured or unstructured. The difference is basically how involved the moderator (the person leading the group) is in the discussion, that is, how active he or she is in asking questions and steering the discussion. In a more structured focus group the moderator follows a previously prepared interview guide and the participants are asked to react to the questions in the guide. In an unstructured focus group the moderator leaves the development of the discussion to the participants and only slightly controls the discussion by keeping them on subject and helping them along when they get stuck. It is advisable that an interview guide should be developed regardless of whether the research group has decided upon a structured or an unstructured approach. Creating an interview guide will help the researchers to structure their thoughts and connect the purpose of the study with the set of question that they would want the participants of the focus groups to discuss. Even in an unstructured focus group it can be important that the moderator has an interview guide to fall back upon if the discussion does not develop in a satisfactory manner. A guide also helps the moderator when it comes to keeping an overall perspective on the discussion.

How many focus groups are needed?

This is a common question but very difficult to answer. The standard answer is that the purpose of the project gives signals on the number of focus groups that are needed, but the final decision is left up to the researcher. A general formula is; when the cost of conducting one more focus group outweighs the gain from that group, it is time to stop. This usually is something the researchers will feel during the process of conducting focus groups, after a while the same arguments will be repeated and the researcher can almost predict what is going to be said. The situation s described as 'saturation'. Our experience is that it is difficult to say, with any certainty, how many groups will be needed before the project is up and running. So we would recommend some flexibility here to adapt to the situation. One important lesson to be learned is not to be overly ambitious in the beginning. It generally takes much more time to plan, conduct and analyse a focus group then is first expected. If the project depends on a large amount of focus groups the project timetable should reflect this, otherwise the researcher will run the risk of not completing the project on time. In River Dialogue 7-9 focus groups were conducted in each country and this was felt to be sufficient.

Focus group results

Are then focus groups a good way of improving and increasing public participation in water management? River Dialogue results indicate that they can provide a useful method for initiating public discussions into complex matters, and then participants can discuss relatively complex matters in them. However, they do not necessarily provide the impetus for further action and empowerment, or for the further gathering of more information. The project strategy was that focus groups should be combined in a two-step process with another method, where the focus groups were used to instigate the process and to gain insights into the ways that water management problems are conceived and constructed in public discourses. The project recommendation is that other methodologies should be utilised to refine insights and to formulate recommendations. In the case of River Dialogue this other method was citizen juries.

The citizens' juries in River Dialogue

The second method that was tested in the River Dialogue project was the citizens' jury. Our discussion here is based on project reports from the use of the method in Estonia (Säre and Unt 2003), Sweden (Gooch and Linderoth 2004) and the Netherlands (Huitema *et al.* 2004) respectively and on a comparison of the three juries (Huitema 2005). The citizens' jury method and the related 'planning cell' method were developed in the early 1970s, by Ned Crosby in the USA and Peter Dienel in Germany respectively. A citizens' jury consists of a group of 12 to 24 randomly selected citizens who attend a series of meetings in order to learn about and discuss a specific issue and make their recommendations public (Crosby 1995). The issue is usually raised in the jury in the form of a question which is called 'the charge' of the jury. With trained moderators ensuring fair proceedings, the jurors hear evidence, question witnesses, discuss the issues that are raised, and then make an informed judgment with regard to the charge (Kuper 1996). The witnesses are selected on the basis of their expertise and/or on the grounds that they represent affected interests (Smith and Wales 2000).

In terms of the authorities assigned to the participants, the jury verdict often has the character of a recommendation to the commissioning body that responds in public to the recommendations made. Jurors are normally allowed to alter the agenda of the jury somewhat, but the broad agenda tends to be determined by the initiator and/or the commissioning body. Usually, the responsible public authority is committed (obliged) to give a response to the recommendation. The group decision rules assume a fundamental assumption that every member of the jury is equal and that the jury tries to take a decision based on consensus. The moderators of the jury are given an important role to impose equality – a difficult task to start with, as it requires a lot of social skill. Most authors of the citizens' jury model emphasize the importance of having a representative group of citizens who draw on a wide range of experiences and backgrounds (Kenyon *et al.* 2003). A citizens' jury is characterised by multiple flows of information and usually includes one or more preparation meetings, in which the jurors meet and receive information on the charge. The actual jury process includes several witnesses (e.g. scientists, other stakeholders) who are selected with the help of the jurors, and who are to reflect various different interests. These witnesses give presentations and are questioned by the jurors, sometimes individually, sometimes with more than one witness at the same time, which gives the jurors the opportunity to confront the witnesses with each others' claims and viewpoints.

Looking at the question of who should participate, it can be seen that because the jury members need to volunteer for the task themselves, certain groups tend to be over represented and in the River Dialogue project this overrepresentation took the form of men over 30 years of age (Estonia and the Netherlands) and men over 40 years of age (Sweden). This type of overrepresentation can be compensated for by introducing minimal quotas for various types of jurors (e.g. no less than 40 percent women, etc.). It appears that the criteria by which jurors are selected differ per country. In the UK, the criterion of social class is often used, whereas in the USA, the criterion of race is used. In a broader sense, it is difficult to determine which community the jurors need to be representative of. Citizens' juries are quite demanding for the participants in the sense that they require preparation, tend to last several days, and demand intensive concentration. Due to this, it needs more effort to commit participants to a citizens' jury than, for example, to a focus group. Furthermore, commitment may be improved by signing a contract with the jurors. Moreover, some additional measures can be taken, such as the provision of day-care for children and financial compensation for lost wages. It appears that, in practice, the percentage of people who are willing to participate in a citizens' jury varies from 5 to 40 percent. In River Dialogue, no financial incentives were provided apart from board and lodging during the days that the jury met. The number of positive replies to the invitations to participate in the juries differed from 6.1% in Estonia to 3.3% in Sweden and 3% in the Netherlands.

Concerning information, in citizens' juries, scientific information plays a prominent role. A citizens' jury includes witnesses (e.g. scientists, NGO's, managers etc.) who provide presentations and who are selected with the help of the jurors. The witnesses preferably represent different views, making clear areas of disagreement and uncertainty that can be questioned by the jurors. A jury is after all designed to encourage an intensive exchange of argumentation. This is both the case during or after the presentations of the witnesses (they are questioned by jurors) and for the process of reaching a final recommendation. This recommendation is normally reached after at least an entire day of group deliberations and often contains motivations. When the jury members disagree (this is fully acceptable) a vote may be taken and 'voting blocks' may motivate their preferences. Jurors may also include reservations in the recommendation.

Organising the River Dialogue juries

As noted earlier the project methodology included both focus groups and citizen juries. The focus groups helped to identify the most relevant issues for water management and also provided some of the experts for the citizen juries. The first step in the process was to recruit the jurors and experts. Different interests were asked to act as experts; these included representatives for local environmental authority, industry, environmental NGO's, farmer's association, fishermen's association, scientific experts etc. The jurors were recruited in slightly different ways.

The jurors

In Sweden a selection of 1500 addresses of people between 18-75, with an equal number of men and woman, was acquired from a statistical institute. Invitation letters with brief details of what was required from jurors and a description of the River Dialogue project were sent out and respondents were asked questions regarding their age, gender, highest education level and profession. 47 answers were received within six weeks after sending out invitations, a response rate of 3,3%. The respondents were phoned and asked about their motivation, awareness about the time investment, and if they were or had been involved in the city council or water management, and 12 jurors were selected. Selected jurors and four reserves were invited to an informational pre-meeting. Due to circumstances all four reserves were accepted on the Jury. In Estonia 800 letters were distributed with local newspapers. The letter contained: a) a description of the Citizens' Jury and an invitation to participate in it; b) a questionnaire on environmental issues in the region and an empty pre-paid return envelope for responses. All together 49 positive answers were received, a response rate 6,1%. The wish to participate might not be too high because of the fact that the tradition of public participation in Estonia was very new and people might have felt estranged when receiving an invitation to the Citizens' Jury, as they did not know what it was all about. Out of the 49 respondent, 15 people were selected, an equal number of men and women, including people with university, gymnasium and primary education, from different professions (teacher, tractor-driver, pensioner, museum-worker, unemployed, NGO person etc.), and also from different age groups. Telephone interviews were conducted with the selected 15 jury members in order to gain more information about their educational-professional background and motivation to participate. In the Netherlands a selection of 2000 addresses in the region received a letter, a reply-card and a flyer. After six weeks 56 positive reactions had been received, giving a response rate of 3.05%. Phone conversations with respondents clarified their motivation to participate and led to the selection of the most suitable candidates. Jurors were then selected to reflect the population of the region (50% men and 50% women). Fourteen jury members, who promised to be present on the pre-meetings and on the Jury days, signed contracts.

Pre-meetings

A few weeks before the jury took place all participants were invited to one or two pre-meetings where the aims of the project were explained and the format of the citizen jury explained. At this meeting the jurors had the chance to ask questions and to provide feed-back on the choice of experts, and on the 'charge' of the jury (see above). The jurors were also asked to sign the citizen jury contract, which stipulated the roles of the organisers and of the jurors.

The jury

The juries were organised over two days in each country and witnesses, 5 in Estonia, 9 in Sweden and 11 in the Netherlands, provided information and viewpoints. Witnesses had about 20 minutes each for their presentations, after that the jurors could discuss with each other in smaller groups of 3-6 persons before gathering again to ask the witnesses questions. At the end of the 2 days, or, in the case of the Swedish jury the following weekend, the jurors formulated their recommendations.

Results

The *overall conclusions* of the three organising teams were positive. Each of the three organising teams reported high satisfaction grades from the jurors and from the witnesses. Every jury contained ordinary members of the public who normally do not get involved in policy making, yet they showed a surprisingly large capability to master relatively complex issues and a fresh perspective on the issues at hand. While the amount of time that each juror spoke differed considerably, all jurors did participate actively, probably thanks to the active roles of the moderators, a role that is extremely important. The organisation of the citizen juries was certainly time and labour-consuming in each of the three countries, yet if the jury's recommendations are incorporated into the water policy process, then the costs in the form of time and expenses may be motivated, especially as the citizen jury method may improve democratic procedures. If however the jury's recommendations are not taken into account then perhaps other, less costly, methods should be considered.

3.3.3 HarmoniCOP[1]

The EU financed project "Harmonising Collaborative Planning" (HarmoniCOP) was designed to improve public participation in River Basin Management Planning (RBMP). The project involved fifteen partners from Belgium, France, Germany, U.K., Hungary, Italy, Netherlands, Spain and Switzerland, and the project's objective was to generate information about public participation in RBMP. The project was supported by external advisers constituting a stakeholder platform, and representatives from NGOs, local government, policy making, water industry and farming were involved in the project as external advisers during its implementation. The project began in November 2002 and ended in October 2005. An important project output was a practical handbook which aimed to address the needs of practitioners in RBMP and to help increase practitioners' understanding of participatory river basin management planning in Europe.

HarmoniCOP was based on the use of nine case studies carried out in nine European countries, which were used as a means of gaining a deeper understanding of participatory processes in Europe. A major aim of the project was to achieve *'an improved understanding of processes of social learning is a prerequisite to implement novel management strategies and to bridge the science-policy gap'* (HarmoniCOP homepage).

A central concept for HarmoniCOP was *social learning*, which 'emphasises *collaboration between the different stakeholders, starting at the earliest possible moment. It helps to build up trust, develop a common view on the issues at stake ,resolve conflicts and arrive at joint solutions that are technically sound and actually implemented in practice'* (Ridder *et al.* 2005). The project examined the potential for social learning processes in the countries involved and concluded that this depended on socio-economic and governing systems (report from HarmoniCOP WP6).

A major result from the project was the production of the handbook 'Learning Together to Manage Together' (ibid.) which provides an excellent overview of both the problems and possible solutions of public participation.

The public participation processes in the HarmoniCop European river basins have been described in detail elsewhere (Mostert *et al.* 2006; Huitema *et al.* submitted)). In several cases relations between different stakeholders improved. In some cases, e.g. the Dordogne basin, a new basin-wide public organisation was established. In most cases many stakeholders got a better understanding of the management issues at stake and got to know and appreciate each other's perspectives. This opened up possibilities for win-win solutions. In several cases the participatory process resulted in clearly identifiable improvements for the stakeholders and for the environment. In the Dee Basin (Scotland), an initial proposal for a wastewater treatment plant had been put forward by local authorities to ensure compliance with the Urban Waste Water Treatment Directive (91/271/EEC). Due to the contentious nature of the proposal, the local community was invited to become involved in the process. This resulted in the development of solutions that the authorities had not previously considered, such as the inclusion of wetlands. As a result, the initial proposal was changed, enabling an increase in amenity values, water quality and biodiversity within the area and greater ownership of the solutions developed. Despite these positive experiences, difficulties have occurred in the HarmoniCOP basins. The single most important difficulty was the lack of clarity about the role of stakeholder involvement. In half of the cases the status of the initiative in which the stakeholders could become involved was not very clear. Often, the organisers lacked decision-making powers. More often than not the stakeholders doubted that their input would make a difference. This can be lethal for the motivation to participate. This lack of clarity about the role of stakeholder participation was only partly a matter of limited and miscommunication. At least as important were political and institutional factors. Quite often the existing governance style was not participatory, and it took a lot of convincing to move towards a more collaborative approach. In many cases the authorities lacked experience with multi-party approaches, relied heavily on

[1] This text is taken from Huitema *et al.*, submitted. We are indebted to Erik Mostert (Delft University) for allowing us to replicate his summary of the HarmoniCOP findings. The reader is referred to Mostert *et al.* 2006 for more information.

technical expertise, were not willing to change, feared to loose control or feared that too broad participation could threaten the confidentiality of proceedings. Consequently, participation often remained limited to information provision or consultation.

Other difficulties included limited resources of organisers for organising participatory processes and of other stakeholders for participating in these processes. This can also result in unrepresentative participation. Typically, stakeholder groups and individuals with much more resources (information, money, time, skills, etc.) are overrepresented and can exert more influence. Unless underprivileged stakeholders are actively supported, PP may actually reinforce power imbalances rather than reduce them. Finally, some processes described in the HarmoniCOP project were plagued by an exclusively technical focus. Some processes took as their starting point technical models instead of the issues as seen by the stakeholders. In some cases overly technical language and overly complex IC-Tools were used, with little attention to communication and interaction between the stakeholders. These difficulties are common to many case studies besides the HarmoniCOP project. They intersect with a fundamental issue: the relationship between PP and the theory of representative democracy. It is commonly argued that the highest level of PP is decision-making by the stakeholders themselves. Representative democracy, on the other hand, would imply decision-making by elected representatives. But it has to be recognised that in practice there are very few decisions that are made by elected representatives only. Most decisions are prepared and influenced by government bureaucracies and government experts and negotiated with well-organised interest groups. PP could be a means to open up this process and make it more transparent. Besides, there are alternative theories of democracy, which stress competition between different interest groups (e.g. Dahl 1971), direct participation by individual citizens (Pateman 1970; Budge 1996) and decentralisation and the use of modern IC-technology.

In many cases implementing PP requires political, institutional and cultural change. Sometimes opportunities for truly participatory approaches may arise at the local level or in specific policy processes - an influential politician may for instance favour PP, or there is a public controversy that cannot be resolved without involving the public. Provided that these processes are well organised, they increase positive experiences with and support for PP. This in turn may prepare the ground for institutional and cultural change (cf. Pahl-Wostl *et al.* 2007).

3.4 CONCLUSIONS

Besides the projects described above, there have of course been many other EU financed projects in which public participation has been part of the project tasks. An overview of some of these projects conducted by the NeWater project (Blind *et al.* 2005) attempted to identify which actors had been involved in participatory processes in the projects, and in which ways. While the results of the study should only be seen as an indication of the degree of participation, they do help to demonstrate that relatively few projects managed to achieve active participation of all stakeholders. This overview of the EU research projects shows that local and regional government are included as important actors in the projects, while national government seems less important. Government was also seen as an important actor in an earlier overview of water management literature conducted in the MANTRA-East project (Höglund and Gooch 2002). Regional government seems to be considered especially important by the projects, probably because in many countries it is this level of administration that water and environmental issues are practically implemented. This emphasis on implementation at the local level and the dependence on different types of actors accentuate the importance of networks. The national authorities, as well as the international agreements, are dependent on the compliance of the actors they cannot force by traditional government measures. Networks play a crucial role in governance process, as there is no actor capable of forcing all other actors to comply. Actors also play a crucial role in the dissemination of knowledge. Knowledge of the impact of pollution can influence how governments and other actors perceive their interest. While there are of course methodological problems involved in the division of government into local, regional and national, as well as in the distinction between operational and policy aspects, it does seem that the local and regional levels play a major role. National governments play, of course, an important role in transboundary water policy, such as in cooperation on the Rhine (Dieperink 1998), yet many other actors play important roles, even in such conditions (Nachtnebel 2000). How can we explain this? First, from the traditional three levels of government and administration, national, regional and local, the development of EU has added a fourth level of governance. The importance of the different levels varies among the European countries, but the influence of the EU, and international, level has increased during the last decades of the 1900's (Gooch 2000), although not always in ways that we would expect. Pierre and Peters argue that states' possibilities to govern have changed since they've become more interdependent, partly due to the process of globalisation which has strengthened the global level of governance, and international organisations like WTO and OECD are growing in importance (Pierre and Peters 2000). In the field of water management there are national laws, EU directives, the Helsinki convention and other international conventions. The implementation of all these laws have to be done on the local level, it is on the local level that the practical work can be done. Aynsley Kellow puts it this way: 'International policy must first be given effect by domestic legislation, and then that policy must be implemented' (Kellow 1997). In an analysis of water management we might consider five spatial levels of decision-making/implementation, the global level, the EU level, the national level, the regional level, and the local level. Implementation occurs at the regional and local levels, and this can probably explain the perceived importance of these levels of government in the project survey.

We have seen in this chapter that public participation is important, but that the ways of increasing participation need to be more clearly defined, and that the aims and ambitions of water managers and authorities need to be clearly specified. We also need to clearly stipulate the ways that the output of the participatory process will be incorporated into management and policy processes. Here we touch one of the major problems of participatory processes in systems of representative democracy. The participants in participatory processes such as citizen juries may be able to raise their own levels of awareness and knowledge, and to formulate coherent recommendations, but they are only able to influence the policy process to the degree that elected politicians and employed administrators are prepared to accept their recommendations. While participatory methods may succeed in providing the informed views of a selection of citizens, and succeed in producing recommendations that have the potential to contribute to the quality of the decision-making, a major issue is still how and if these recommendations will be acted upon. If participatory processes are to be developed to actively involve stakeholders and the public in the implementation of water policy then mechanisms must also be created that not only facilitate participation, but that also allow the views and interests of these groups to be included into the decision-making and policy processes that determine the scope and outcomes of water management.

To conclude, more knowledge and experience of the forms of participation that are suitable for different aims and for use with different groups (citizens, effected citizens, stakeholders etc) needs to be developed. It should be clear by now that in this chapter we have hoped to contribute to the further shaping of participatory practices in water management; however, we fully realise that devising and recommending participatory processes also implies choices and preferences, such as who gets to participate and who not, in what way, and who gets to represents whom (Young 1990). These are issues that will need to be addressed in the coming years if the promise of improved participation in water management is to be realised.

3.5 REFERENCES

Arnstein, S. (1969) A ladder of citizen participation. *Journal of the American Institute of Planners.* **35**(4): 216-224.

Arts, B. (2000) Regimes, Non-State Actors and the State System: A "Structurational" Regime Model. *European Journal of International Relations* **6**(4): 513-542.

Birch, A. H. (1993) *The concepts and theories of modern democracy.* London, Routledge.

Blind, M., F. Hattermann, J.G. Timmerman and B. Sikorska (2005) European IWRM – Research outline analysis of projects' contribution for the purpose of selecting topics for synthesis; A discussion starter to the European IWRM platform. Report No 3 of the NeWater project RIZA, The Netherlands.

Budge, I. (1996) *The new challenge of direct democracy.* Cambridge, MA: Polity Press.

Bulkeley H,A.P. and J. Mol (2003) Participation and environmental governance. Consensus, ambivalence and debate. *Environmental Values* **12**: 143 - 154.

Coenen, F., D. Huitema, et al., Eds. (1998) *Participation and the quality of environmental decision making.* Dordrecht, Kluwer Academic Publishers.

Cohen, J. and A. Arato (1994) *Civil Society and Political Theory.* Cambridge, MA, MIT Press.

Crosby, N. (1995) Citizens juries: One solution for difficult environmental questions. In: O. Renn, T. Webler and P. Wiedemann (Eds.) *Fairness and competence in citizen participation. Evaluating new models for environmental discourse.* Dordrecht, Kluwer Academic Publishers: 157-174.

Crosby, N. (1998) Using the citizens' jury process for environmental decision making. In: K. Sexton, A.A. Marcus, K.W. Easter and T.D. Burkhardt (Eds.) *Better Environmental Decisions. Strategies for Governments, Businesses and Communities.* Washington, DC, Island Press.

Dahl, R.A. (1971) *Polyarchy: participation and opposition.* New Haven: Yale University Press.

Dalton, J.R. (2004) *Democratic challenges, democratic choices, The erosion of political support in advanced industrial democracies.* Oxford, Oxford University Press.

Dieperink, C. (1998) From open sewer to salmon run: lessons from the Rhine water quality regime. *Water Policy* **1**(5): 471-485.

Downs, A. (1972) Up and Down with Ecology - the "issue-attention cycle". *The Public Interest* **28**: 38-50.

Dryzek, J.S. and C. List (2003) Social Choice Theory and Deliberative Democracy: A Reconciliation. *British Journal of Political Science* **33**(1): 1-28.

EU (2000) Community legislation: Directive 2000/60/EC of the European Parliament and of the Council of 23 October 2000 establishing a framework for Community action in the field of water policy. Retrieved 2001-06-20, 2001, from http://europa.eu.int/eur-lex/en/lif/dat/2000/en_300L0060.html.

Gastil, J. (2000) Is Face-to-Face Citizen Deliberation a Luxury or a Necessity? *Political Communication* **17**: 357-361.

Gellner, E. (1995) The Importance of Being Modular. *Civil Society*. J.A. Hall. Cambridge, Polity Press.

Giddens, A. (1991) *Modernity and Self-Identity in the Late Modern Age*. Cambridge, Cambridge University Press.

Gooch, G.D. (2000) *East is east and west: municipal co-operation and regional networks around the Gotha canal*. Linköping, Linköping Universitet.

Gooch, G.D. (2003) *Scenarios for Environmental Policy-Making on the Estonian-Russian Border*. Linköping, Department of Management and Economics, Linköpings University: 17.

Gooch, G.D. (2006) From Dialogue to Trialogue: Sustainable Ecosystem Governance and Civil Society. In: A.R. Turton, D. Roux, M. Claassen and J. Hattingh (Eds.) *Governance as a Trialogue: Government-Society-Science in Transition*. Berlin, Springer-Verlag.

Gooch, G.D. and C. Linderoth (2004) *Report on the Swedish citizens' jury of the River Dialogue project*. Citizen involvement in exploring the future of the Motala Stroms basin river area. Linköping, Linköpings University.

Gooch, G.D. and P. Stålnacke, Eds. (2006) *Integrated Transboundary Water Management in Theory and Practice: Experiences from the New EU Eastern borders*. London, IWA Publishing.

Governance, T. C. o. G. (1995) *Our Global Neighbourhood*. Oxford, Oxford University Press.

Hall, J.A. (1995) In Search of Civil Society. *Civil Society*. J.A. Hall. Cambridge, Polity Press: 1-31.

Höglund, P. and G.D. Gooch (2002) *Review of Existing Literature in Transboundary Water Management*. Linköping, Linköping School of Management.

Huitema, D. (2005) A comparative analysis of the three citizens' jury's under River Dialogue. In: K. Kangur (Ed.) *Focus groups and citizens' juries. River Dialogue Experiences in enhancing public participation in water management*. Tartu, Peipsi Center for Transboundary Cooperation: 44-50.

Huitema, D., M. Van de Kerkhof, et al. (2004) *Exploring the future of the IJsselmeer*. A report to the River Dialogue project on the Dutch Citizens' Jury. Amsterdam, IVM.

Huitema, D., W. Egas, S. Moellenkamp, E. Mostert, C. Pahl-Wostl, and R. Yalcin, (submitted) Adaptive water governance. Assessing adaptive management from a governance perspective. *Ecology and Society* (submitted)

Jouvenel, H. d. (2000) A Brief Methodological Guide to Scenario Building. *Technological Forecasting and Social Change* (65): 37-48.

Jun, J. S. (2002) New Governance in Civil Society Changing Responsibility of Public Administration. In: J.S. Jun (Ed.) *Rethinking Administrative Theory: The Challenge of the New Century*. Westport, CT, Praeger: 289-307.

Kellow, A. (1997) "Problems in international environmental governance" or "A policy analyst looks at the world. this being a tale of how the hopes of mice and men in Geneva are dashed in Canberra". *Australian Journal of Public Administration* **56**(2): 54-65.

Kenyon, W., C. Nevin, et al. (2003) Enhancing Environmental Decision-making Using Citizens' Juries. *Local Environment* **8**(2): 221-232.

Kuper, R. (1996) *Citizens juries: The Hertfordshire experience*. Hatfield, University of Hertfordshire. Business School.

McLaverty, P. (2002) Civil society and democracy. *Contemporary Politics* **8**(4): 303-318.

Mostert, E., C. Pahl-Wostl, Y. Rees, B. Searle, D. Tàbara, and J. Tippett (2006) Social learning in European river basin management; Barriers and supportive mechanisms from 10 river basins. *Ecology and Society* (in press)

Nachtnebel, H.-P. (2000) The Danube river basin environmental programme: plans and actions for a basin wide approach. *Water Policy* **2**(1-2): 113-129.

OECD (2001) *Engaging Citizens in Policy-making*. Puma Briefing Paper. Paris, Organisation for Economic Cooperation and Development.

Pahl-Wostl, C. (2002) Towards sustainability in the water. The importance of human actors and processes of social learning. *Aquatic Sciences* **64**: 394-411.

Pahl-Wostl, C., M. Craps, A. Dewulf, E. Mostert, D. Tabara, and T. Taillieu (2007) Social learning and water resources management. *Ecology and Society* **12**:(2):5

Pateman, C. (1970) *Participation and democratic theory*. Cambridge: Cambridge University Press.

Pierre, J. and B.G. Peters (2000) *Governance, Politics and the State*. Houndmills, Basingstoke and London, Macmillan Press Ltd.

Porter, M. (1985) *Competitive Advantage*. New York, Free Press.

Ridder, D., E. Mostert, et al., Eds. (2005) *Learning together to manage together. Improving participation in water management*. Osnabrück, University of Osnabrück.

Roll, G. and J.G. Timmerman (2006) Communicating with stakeholders and the public. In: G.D. Gooch and P. Stålnacke (Eds.) *Integrated Transboundary Water Management in Theory and Practice: Experiences from the new EU eastern borders*. London, IWA.

Säre, M. and P. Unt (2003) *Report of citizens' jury. Water transport on River Emajógi in the Alam Pedja nature reserve*. What would be the compromise between the interests of environmentalists, entrepreneurs and local inhabitants? Tartu, Peipsi CTC.

Smith, G. and C. Wales (2000) Citizens' Juries and Deliberative Democracy. *Political Studies* **48**(1): 51-65.

Smith, S.R. and H. Ingram (2002) Policy Tools and Democracy. In L. Salamon, (Ed.) *The Tools of Government: A Guide to the New Governance*. New York, Oxford University Press: 565–84.

Van de Kerkhof, M. and D. Huitema (2004) Public participation in river basin management. A methodological perspective. In: J.G. Timmerman, H.W.A. Behrens, F. Bernardini et al. (Eds.) *Information to support sustainable water management. From local to global levels*. Lelystad, RIZA: 141-148.

Wibeck, V. (2000) *Fokusgrupper: om fokuserande gruppintervjuer som undersökningsmetod*. Lund, Studentlitteratur.

Young, I.M. (1990) *Justice and the Politics of Difference*. Princeton, Princeton University Press.

4

Analysis of European IWRM research on transboundary regimes

Gulnara Roll, Natalia Alexeeva, Eduard Interwies and Jos G. Timmerman

4.1 REGIMES: TRANSBOUNDARY INTERACTIONS IN WATER MANAGEMENT

4.1.1 Introduction and topic delineation

There are more than 260 major transboundary rivers in the world (Wolf *et al.* 1999). The increasing water scarcity and depletion of natural resources lead to the increase in water conflicts between countries that share transboundary waters; and geography of such conflicts is expanding (Yoffe *et al.* 2004). To prevent and resolve the conflicts, riparian states need to agree on common rules and procedures of cooperation on the transboundary waters, i.e., to establish transboundary water regimes. Examples of transboundary water regimes are, for instance, the Helsinki Convention on Transboundary Waters (UNECE 1992) and the Estonian – Russian Lake Peipsi/Chudskoe transboundary water agreement.

In order to better understand the challenges and opportunities linked to the management of transboundary waters, it is useful to establish first the basis of international regime theory. International regimes are "implicit or explicit principles, norms, rules and decision-making procedures around which actors' expectations converge in a given area of international relations" (Krasner 1983). The definition of international regimes includes both formal as well as informal procedures and rules of the international cooperation on different levels of governance – from very local to global. Transboundary water regimes is one area of studies of international regimes. Transboundary water regimes usually include both formal rules such as international water conventions, statutes of transboundary water commissions, cooperative agreements adopted by national governments and aimed at coordinating national water management activities in transboundary water basins, relevant national laws and procedures as well as informal rules such as, for an example, traditional ways of using natural resources (traditional ways of transport or fishing, etc.) that are informally accepted in transboundary water basins but are not documented as formal norms in agreements or contracts).

In this chapter we will discuss developments in research on transboundary regimes with a focus on European research. First, we will discuss essential concepts in regime research and how this relates to IWRM and AM. Next, we will discuss how this research was done in Europe and focus on several EU research projects that included transboundary regimes research. Finally, we will conclude how AM is limited in its approach towards regimes and how transboundary regimes research can improve AM.

 The adaptiveness of IWRM; Analysing European IWRM research Edited by Jos G. Timmerman, Claudia Pahl-Wostl and Jörn Möltgen. ISBN: 9781843391722. Published by IWA Publishing, London, UK.

4.1.2 Focus in regimes' research

The study of international regimes has been an important part of research in international relations for over three decades. The term "international regimes" was introduced first in 1975 (Ruggie 1975) and in-depth studies of regimes started in the 1980s (Krasner 1983). Initiation of the international regimes research marked a shift away from an exclusive focus on formal international organisations (Brahm 2005). Studies of international regimes are conducted on three major large topics: (1) formation of international regimes; (2) implementation of regimes, including regime effectiveness; and (3) the regimes fit and scale and regimes' interaction. Short explanations of these three thematic areas of studies of international regimes are presented below.

Research on formation of international regimes

The literature on formation of international regimes focuses on the topics of how actors in international society (nation-states, international organisations, NGOs, scientists and other) are involved in regimes' design and development; what kind of institutional arrangements or regimes they elaborate to cope with transboundary problems; and why those actors succeed in addressing some of the transboundary problems but fail to do so in connection with other, seemingly similar, ones (Keohane, 1984; Young 1989a; Young and Osherenko, 1993; Elliott, 1998; Porter et al 2000). Three stages of regime formation have been identified - agenda formation, negotiation, and operationalisation (Young, 1996).

Methodological approaches to studies of regimes' formation and empirical studies of international regimes in Europe were recently described by Lindemann (2006). One of the case studies described by Lindemann is the formation of the Rhine water basin regime. Rhine River basin is shared by the Netherlands, France, Germany, Luxembourg and Switzerland. The Rhine transboundary water regime formation started with negotiations on establishment of the International Commission for the Protection of the Rhine (ICPR) that was established in 1950. In 1963 the riparian countries signed an Agreement on the international commission for the protection of the Rhine against pollution and in 1976 - the Rhine Chemical and Chloride Convention. The Rhine water regime was operationalised with the adoption of the Rhine Action Programme (RAP) in 1987. Lindemann described a process of negotiations between the riparian countries during the whole period of the regime formation and identified roles and interests of actors involved. According to Lindemann, a driving force behind the creation of the Rhine water regime was the Rhine River downstream country – the Netherlands.

Research on implementation of international regimes and institutions performance

Next step in the life of a regime is its implementation. One set of questions that concerns implementation of regimes is about compliance of actors with the commitments taken - the agreed rules, procedures and tasks. Another set of questions is about performance of international regimes as institutions – how effectively the regimes function; are they sufficiently robust and effective to ensure achieving a goal of a regime?

Issues of the regime implementation and compliance were studied by Chayes and Chayes (1991), Underdal and Hanf (2000), Victor (1998), and Young (1997, 1999, 2001). In a broader sense, international regime effectiveness is related to the literature on public policy evaluation (e.g. Mohr 1988). The regime effectiveness is evaluated from the point of view of achieving a goal of an international regime (e.g. decreasing pollution) through changing behaviour of actors involved in a regime. Another approach to looking at effectiveness of international regimes is from the point of view of regimes long-term sustainability (see more in Young 2001). For instance, in a transboundary water basin one environmental issue may be on an agenda today (e.g. water eutrophication), but in a decade when the earlier existing problem is resolved, a new challenge may emerge. Established cooperative institutions should have sufficient capacity and flexibility to be able to react to new emerging challenges and adjust appropriately to those changes.

A series of empirical studies of measuring regime effectiveness were conducted using expert review teams, simulation exercises, political cost-benefit analysis, statistical effect separation and other methods (see, for instance, Hovi *et al.* 2003). Those studies revealed a quite modest success in achieving commitments and implementation of regimes in general. Therefore, implementation of international regimes is an important topic also for future research.

Research on regimes' fit, scale and interaction

Other important topics in the regimes' research were the regimes' fit and scale, and regimes' interaction. Those studies were developed by Young (2002). The topic of the regimes' fit and scale is important for studying the interaction of different regimes on different governance scales in Europe – for instance the interaction of the Baltic Sea Basin water management regimes with "enclosed" bilateral water management regimes within the Baltic Sea Basin such as transboundary basins of rivers Odra, Narva, Daugava and other.

The regime interaction can take place between regimes that deal with the same topic (such as water quality) but have different geographical boundaries – for instance, the multilateral Baltic Sea and bi-lateral Narva River water management regimes – or international regimes that address issues related to different but interlinked sectors, such as protection of transboundary waters and economic security; or use of shared natural resources and arms control (see, for instance, Chalecki and Gleick 2001). Young (2002) stressed the importance of considering those intersectoral links while studying regimes: "*at the international level, environmental challenges constitute only a small segment of the overall agenda discussed by stakeholders, which also encompasses institutional, security, economic and other issues*".

4.1.3 Different theoretical perspectives to studies of international regimes

Despite the fact that research on international regimes is conducted for about three decades now, scholars have never arrived to one commonly accepted concept of regimes formation and development. Hasenclever (1997) identified three major theoretical schools of international relations – realist, neoliberal and cognitive - that offer different explanations for the origins of regimes and their relative influence (see a summary in Table 1). Lindemann (2006) proposed to add to the mentioned three concepts one more - what he names a contextual approach, an approach that allows to explain emergence of international regimes in a context of „national and international events that are seemingly unrelated to the issue under consideration" that basically trigger establishment of international water regimes. We will not consider this fourth concept in this chapter, as it describes the trigger that accelerates the establishment of a regime rather than providing an additional concept on the conditions and reasons for regime building.

Table 1: Theoretical Perspectives on International Regimes (Source: Hasenclever *et al.* 1997)

	Realism	Neoliberalism	Cognitivism
Central Variable	POWER	INTERESTS	KNOWLEDGE
Metatheoretical Orientation	RATIONALIST	RATIONALIST	SOCIOLOGICAL
Behavioural Model	RELATIVE GAINS SEEKER	ABSOLUTE GAINS MAXIMISER	ROLE PLAYER
Institutionalism	WEAK	MEDIUM	STRONG

Realism

The theoretical school of realism directs our attention to the role of power in creating and sustaining regimes as well as what consequences regimes may have for the distribution of power in the international system. Realists contend that the nature of anarchy in the international system causes states to be primarily concerned about relative gains, i.e. their position vis-à-vis other states in the system. As a result, states will be reluctant to enter into any agreement that leaves them in a worse position relative to others, whether due to the distributional consequences or due to the costs of maintaining the regime.

One of the issues that the realist school addresses (which is an important issue for the discussion of management of transboundary waters) is states' sensitivity to relative gains, i.e. that others may get more out of an agreement than I do (Grieco 1990, 1993). States fear suffering relative losses. Because states are interested in security, their sensitivity to relative losses varies over time and across issue area. Thus, cooperation will be more likely in economic areas than in military ones, concludes Grieco.

Neoliberalism

The second school of neoliberalism views regimes as mechanisms that facilitate achieving optimal outcomes by reducing uncertainty. Neoliberals start from the same assumptions as realists as to the anarchic nature of the international system. However, in contrast to realists, they consider that states are concerned primarily with absolute gains. As a result, when deciding whether to cooperate, states will evaluate what is in it for them, rather than how they will come out relative to others. Therefore, the main concern for states is whether they are getting the best deal possible. Early work on regimes from a neoliberal perspective argued that regimes were created to solve Prisoner's Dilemma situations where states have a common interest in cooperation (Keohane 1984). Regimes can facilitate cooperation by providing information, reducing transaction costs, facilitating linkages, and lengthening "the shadow of the future." Oran Young (1989b) argued that those early regime studies have been too optimistic about reaching an agreement because they do not properly account for the uncertainty actors feel about their strategy and that of others. For this reason, states focus on the bargaining process itself rather than distributive issues, which make cooperation more likely. Young created an institutional bargaining model of regime formation.

Neoliberals have provided important insights into how regimes can facilitate cooperation amongst states. However, a number of shortcomings remained. Aside from those mentioned above, they often underestimated the fear of cheating that states feel. A contrasting criticism came from a cognitive perspective: neoliberals did not address the sociological dimension of regimes; regimes not only facilitate cooperation, but change perceptions of oneself and one's interests.

Cognitivism

The school of cognitivism draws attention to the fact that regimes are fundamentally social entities. Therefore, norms, identities and discourse are important in shaping regimes and are, in turn, themselves influenced by regimes. Cognitivists' contributions to our understanding of regimes begin with an understanding that actors' behaviour is not shaped so much by material interest, but by their role in society. According to Hasenclever *et al.* (1997), unsatisfied with rationalist (realist or institutionalist) theories of international politics, cognitivists generally hold that knowledge and values do not only affect power and shape interests but also play a more independent role in the formation of international regimes. Hasenclever *et al.* divided cognitivists into weak and strong versions. Weak cognitivists are interested in exploring the influence of ideas on actors (Goldstein and Keohane 1993; Haas 1992). Ideas serve to reduce uncertainty and as a means of learning. They serve as road maps, institutions, and focal points. Learning may alter strategies or even goals. In such an environment, consensual knowledge and epistemic communities facilitate policy innovation, diffusion, selection, and persistence. In this view, regimes may result from an idea gaining prominence.

Beginning from an assertion that the international system is fundamentally a social one, strong cognitivists argue that this social structure constructs actors' identities (Cox 1986). Generally speaking, strong cognitivists seek to theorise about the sense of obligation that exists and what accounts for the varied pull of compliance. Regimes construct identities by delineating what are socially acceptable norms and interests.

Adaptiveness in regimes

As a part of the cognitive approach to studies of international regimes, a concept of adaptive regimes was introduced in Chapter 1 and further discussed in Chapter 2 of this book. This concept is related to topics of studies of flexibility and adaptability of international regimes to changing conditions. The concept of adaptive environmental management originates from systems and complexity theory and was introduced in the 1970s (e.g. Holling 1978). In recent years the interest in the application of the concept has increased strongly (e.g. Geldof 1995; Pagan and Crase 2004; Pahl-Wostl 2004; Tompkins and Adger 2004). Geldof (1994) identifies five elements of (active) adaptive water management: "humanising" water management (communication); flexibility of evaluation mechanisms; learning to manage complexity; accepting subjectivity; and accepting uncertainty to a certain level. A process of learning of actors involved in the regimes' implementation is in a centre of attention of scholars of "adaptive regimes' management". Three ways can be identified to structure management as an adaptive process (Pagan and Crase, 2004): evolutionary or trial and error management, passive AM (using lessons from the past) and active AM (using policy and its implementation as a tool for accelerated learning). The latter is adopted as the preferred approach in Chapters 1 and 2.

Turning back to the three theoretical perspectives as discussed above, it becomes clear from this short discussion that knowledge plays a central role in an adaptive regime through its focus on communication, evaluation complexity and uncertainty. The "humanising" and learning perspectives point at a sociological orientation, while accepting subjectivity points at accepting the role-playing of actors. Finally, as discussed in Chapter 1, an adaptive regime that adopts active AM is considered to institutionalise learning. The school of cognitivists therefore best covers an adaptive regime. In our review of the research on transboundary regimes we will consider several projects that further the concept of the cognitivist school of regimes as building towards AM and will identify the shortcomings in this approach.

4.2 RESEARCH ON TRANSBOUNDARY REGIMES IN EUROPE

4.2.1 Brief history and patterns of studies of international regimes in Europe

Initially American scholars of international relations – mostly economists and political scientists, have elaborated the theories of international regimes. In Europe development of the three perspectives to studying international regimes as described above – realism, neoliberalism and cognitivism – have different destinies in terms of the scholarly development in Europe. Authors agree with Lindemann's (2006) conclusion that there is no "one-answer-fits-all" – in different political and geographical contexts, different conceptual approaches showed to be useful in explaining the formation and implementation of international water regimes. For instance, on transboundary waters shared by the EU member states (Germany and the Netherlands, etc.) between whom borders are "melting", there are no major institutional barriers, communication across borders could be organised very easily and the cognitive approach to studies describes better the functioning of international regimes. At the same time, communication between stakeholders takes place as long as there is trust between partners across the borders. When a conflict situation arises between riparian countries that brings distrust to the information coming from the other side of the border (as for instance, the experimental study conducted by Timmerman and Langaas demonstrated – see more on the study below), channels for the information exchanges and cross-border learning get shut down. Then the neoliberal or realist concepts explain better the development and implementation of international regimes. Therefore, all the different perspectives are worth to be discussed. It should be noted that the realism perspective did not really receive further development in works of European scholars. The developments of the neoliberal and cognitivist concepts are described below.

Development of a neoliberal concept of regimes' research in Europe

The neoliberal concept of international regimes received development in Europe thanks to the large-scale international research cooperation projects implemented at IIASA - International Institute for Applied Systems Analysis (Vienna, Austria). IIASA is a special international research organisation established jointly by American, European and Soviet governments after the Second World War and during the Cold War as a way of 'building a common future' of the three large regions.

The project on Implementation and Effectiveness of International Environmental Commitments (IEC) was implemented at IIASA in 1994 – 1997. The primary research focus was on the history of existing agreements and developments in new at the time international environmental agreements. The work on the project was organised around a common research framework to answer three clusters of questions. 1) What factors affect participation in the implementation of environmental agreements, and how do different patterns of participation influence the effectiveness of international agreements? 2) How has the transition in the formerly centrally planned economies, especially Russia, influenced implementation of international environmental agreements; and, how can factors such as the stabilisation of property rights and transfer of resources help to improve implementation? 3) How has the presence of mechanisms for reviewing national implementation and compliance promoted the effectiveness of international agreements? The IEC project was completed in 1997. The International Regimes Database (IRD) that has been developed within the project was an important methodological innovation that allowed scholars to adopt a quantitative approach to the study of international regimes (www.iiasa.ac.at/Research/IEC) (Breitmeier *et al.* 2006; Victor *et al.* 1998).

The international research cooperation initiated within the IEC project, continued further within a global research programme on the Institutional Dimensions of Global Environmental Change (IDGEC) (http://fiesta.bren.ucsb.edu/~idgec/). This included studies conducted in EU research projects such as the EU FP4 research project "International Governmental Organisations and National Participation in Environmental Regimes: the Organisational Components of the Acidification Regime" that will be discussed later on. The major focus of the IDGEC programme was on elaboration of approaches to measuring the effectiveness of international environmental regimes (Sprinz 2000; Helm and Sprinz 2000). As a part of the programme, the UK Global Environmental Change Programme focused on studies of regimes scales; on the issues of management of the linkages between levels, and on global environmental regimes, national governments and local communities (ESRC 2000). Both IEC and IDGEC focus on interests and absolute gains maximising and are therefore included in the neoliberalist school.

Independently from the IDGEC network, scientists at the UNESCO Center for Water Law, Policy and Science at Dundee University, UK, took up the neoliberal approach to studies of international regimes. The Centre research agenda includes studies of water rights/ legal entitlements, frameworks for water allocation and institutional mechanisms as the core-stones of international waters regimes. Particularly, the research covers the regime building (Kibaroglu 2002) outlining demands for water resources and economic considerations as drivers and revealing the changes in institutional settings; role of international institutions such as the UN (Wouters *et al* 2005, Wouters forthcoming). While the IDGEC network of scholars includes into its definition of regimes all the formal and information institutions, including laws and regulations; the Dundee University researchers define laws and institutions as two separate elements of international regimes (Wouters forthcoming). An obvious explanation for this difference in the definitions is that while the "IDGEC network" is led by political scientists, the Dundee University researchers are lawyers by training.

In many ways, as a result of the IEC and IDGEC programmes, an international community of scholars of international regimes studies in Europe was established; the regimes studies were established as a research field. The neoliberal stream of the regime research contributed greatly to the methodology of studies of international regimes, including transboundary water, regimes formation and implementation; especially the quantitative approaches to measuring effectiveness of regimes.

For the future development of the regimes' research in Europe, it is very valuable that within the IDGEC the Programme Science Plan: Institutional Dimensions of Global Environmental Change (2005) has been compiled. The Science Plan is a document that informs about the research conducted earlier within the IDGEC programme and outlines perspectives for future international regimes research. The Science Plan aims at consolidating the regimes research. It delineates "disciplinary" borders for the interdisciplinary field of research of international regimes. Existence of the Science Plan supports long-term sustainability and high quality of national and international research of international regimes. A methodology underlying the IDGEC Science Plan in overall reflects the neoliberal stream of regime research however the Science Plan is constructed in an open and inclusive manner so that also topics and approaches to studies of international regimes used in rationalist or cognitivist schools of thought could be further developed in a framework of the Science Plan. For instance, social learning, a central topic of regime studies in the cognitivist school, is one of important topics addressed in the Science Plan.

Development of a cognitive concept of regimes in Europe

The cognitive approach to studies of international environmental regimes, especially the weak cognitivism that explores the influence of ideas on actors, received much attention among European scientists as a response to a need to study impacts of the increasing in the era of globalisation information flows and use of new communication technologies to the management of transboundary natural resources.

Swedish political scientists Lidskog and Sunqvist studied the role of scientific knowledge in the functioning of environmental regimes (2002). Three key findings are central to understanding the role of science in environmental regimes according to Lidskog and Sunqvist, namely that knowledge never moves freely, that the value of science is the result of negotiations and that science and policy are co-produced. Lidskog and Sunqvist demonstrate their finding using a case study of one of the most science-based regimes - the Convention on Long-Range Transboundary Air Pollution (LRTAP). Lidskog and Sunqvist stress an importance in functioning of environmental regimes of a "dialogue, discussion and deliberation, where social learning . . . is put at the centre" (2004).

A significant contribution to studies of information, knowledge, social learning in water management was made by Dutch water management scientists. Internationally recognised applied water management research has developed in the Netherlands due to an urgent need to address the country's acute water management problems – high risks of floods in areas with high density of population. According to Mostert (2006), the Integrated Water Management approach became a part of the Netherlands' government policy and the IWRM concept emerged in the specific context of the Netherlands water management practices. Research on the IWRM was extensively supported by the Dutch government and water enterprises. It should be also noted that the Dutch water management research focuses a lot on stakeholders, their interests and behaviour; this focus reflects the political tradition of a very interactive and deliberative decision-making system in the Netherlands. Based on the IWRM concept, Dutch water management scientists developed studies of cooperation processes in transboundary water basins and involvement in water management of diverse actors (see for instance Raadgever and Mostert 2005).

Based on the Dutch traditions of the IWRM research as described above and the DPSIR framework (Driving Forces-Pressures-State-Impacts-Responses) that is based on systems analysis, studies on the role of environmental information in managing transboundary waters were developed by Timmerman and Langaas (2005). According to Timmerman and Langaas, information plays an important role in the decision-making, ranging from use of information as a source of power to use of information to postpone decisions to applying information as a basis for cooperation. Timmerman and Langaas focus is on the actors, their actions and behaviour. Timmerman and Langaas search answers to the following questions concerning the actors and learning in their study. Who collects and produces information in transboundary water basins? Who interprets information? Who uses information (if any)? How is the information used in decision-making processes in a transboundary context? Timmerman and Langaas conclude that although the information exchange and communication is an important part of the transboundary water management, information production lags behind developments in water management; and although integrated water management was introduced more than a decade ago, information still focuses on ecological components of water bodies and largely ignores the importance of socio-economic data. Even though better information may be produced, the use of information will still be limited because of: different valuation of information by people with dissimilar beliefs, values and norms, and different cultural habits; insufficient access to information for all actors; insufficient communication channels; and insufficient coordination between the different levels and scales of water management.

Dutch social scientists contributed to studies of management of transboundary waters with a concept of the adaptive water management (van der Brugge and others 2004). Rutger van der Brugge[1] who has a background in environmental health developed a theoretical framework based on the interdisciplinary research paradigm of Complex Adaptive Systems theory (CAS). The CAS theory is related to studies of economics, psychology, biology and cybernetics[2]. Van der Brugge discussed the conditions and role of social learning of actors involved in management of international waters. Transboundary institutions, normally stable in their nature, will adapt (or will not adapt) to new given situations under certain 'pressure' factors.

An adaptive management approach as a method of integrating research, policy and local practices into a structured cycle of learning about changes in water basins, is being developed by a group of scientists in Central Europe (see a.o. Magnuszewski *et al.* 2005 and Sendzimir *et al.* 2006). They are dissatisfied by the traditional environmental management and assessment techniques which "ignore facets of system structure that underlie complexities in water basins" (Magnuszewski *et al.* 2005) and apply the adaptive management approach to river basin management studies. The adaptive management approach is capable in their opinion to address links between social and ecological systems in water basins, and also include perceptions of stakeholders into the analysis and decision process. Similar to van der Brugge, Sendzimir and Magnuszewski try to overcome a monodisciplinary approach that often results in a narrow,

[1] Source: http://www.eur.nl/fsw/staff/homepages/vanderbrugge/

[2] Source: www.wikipedia.org

often very technological focus to water management, and promote an interdisciplinary approach to research and management that reflects the complexities and interdependencies in and between the natural and socio-economic components of water management systems.

It can be concluded that the cognitive school of regimes' research is developing in Europe bottom-up; a number of research groups in the Netherlands, Sweden as well as other European countries use the adaptive management approach to their studies of the management of transboundary water resources. However, those research groups use different methodological frameworks for their studies. Unlike the situation with the neoliberal school of thought that has developed in Europe in a way through an up-down process (the methodological basis was initially elaborated within a long term international research IEC project implemented at IIASA in 1994 – 1997 by a group of scientists from the USA, Europe and Russia and more scholars joined the research network on a later stage through the IDGEC programme), in the authors' opinion, studies in Europe using the cognitive approaches to regimes' that develop bottom-up, have not yet consolidated into one coherent school with a commonly accepted terminology. There is a need in making a more thorough review of those studies and elaborating a common methodological framework for future studies.

4.2.2 EU projects, dealing with transboundary waters

Studies of international regimes until now received a limited attention in EU funded research and technological development projects (RTD). However, the quantity of projects dealing with the regime research is growing. Table 2 gives an overview of a series of EU projects dealing with transboundary water management. Details of the EU projects are given in Table 3. Below, five projects are described where the first two follow the neoliberal perspective in their research and the rest a cognitive perspective to studies of international environmental regimes.

Research project on International Governmental Organisations and National Participation in Environmental Regimes: the Organisational Components in the Acidification Regime

This research project (EU Contract Nr. EV5V-CT94-0390) focused on a methodologically important issue for transboundary water management research of how to assess the international regimes' effectiveness. Methodology of the study was based on the neoliberal approach to regime research. The main conceptual topics that the project addressed, included (see Underdal, 1992): (1) what precisely constitutes the object to be evaluated; (2) Against which standard is the object to be evaluated; (3) How do we operationally go about comparing the object to our standard; in other words, what kind of measurement operations do we perform in order to attribute a certain score of effectiveness to a certain object (regime)?

As a part of this project, Helm and Sprinz (1999) elaborated a standardised measurement concept for the effectiveness of international environmental regimes. They proposed that the actual regime effectiveness is compared, at one hand, to the degree of the instrument use (for instance, percentage of pollution reduction) that would occur in a situation of absence of the international regime under investigation. They call this a "no-regime counterfactual" (NR). At the other hand, effectiveness is compared to the degree of the instrument use in a situation of a "perfect regime" that is named a "collective optimum" (CO). Accordingly, the regime potential is the distance between the NR and the CO, expressed in units of instrument use. Countries will execute actual policies (AP) which fall into this interval. The effectiveness of a regime (E) can then be measured as the relative distance that the actual performance has moved from the no-regime counterfactual towards the collective optimum, or as the percentage of the regime potential that has been achieved (Figure 1). This score falls into the interval (0, 1).

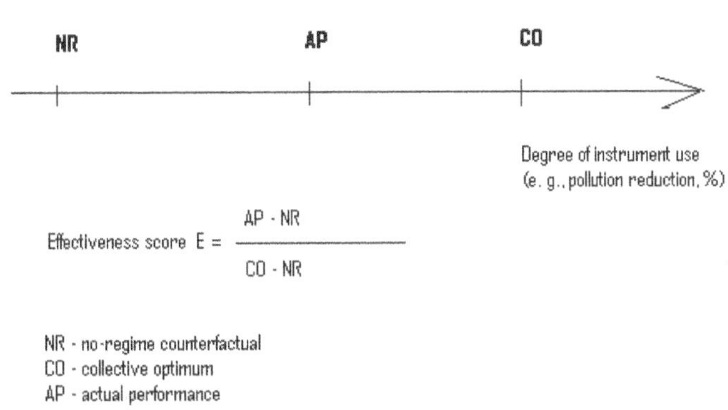

Figure 1: Measuring regime effectiveness – the general concept. Source: Helm and Sprinz (1999).

Table 2: Transboundary aspects in the relevant EU-financed projects

Project acronym/ Transboundary aspects	FP4 National Participation in IER	FP4 Concerted Action	MANTRA-EAST	TRANS CAT	TRANS MAP	TRABO REMA	EURO WATER	RIVER DIALOGUE	HARMONI COP	MPA in EBS	BERNET-CATCH	ADAM
Major addressed issues*	1,2,4, RM, LF, PP	1,2,4, RM, LF, PP	1,2,4, QL, E, EC, GR, RM, PP	1,3 QN, QL, LF, PP	4,6 QL, E, LF, GR	2 E, LF, GR	n/a EC, LF, GR	1,2, GR, PP	n/a PP, GR	4,6, E, RM, GR, PP	1, 2, QL, GR, PP	1, 2 QL, E, M, LF, GR, PP
Transboundary dimension/focus	EU WFD, multi-national	EU WFD, multi-national	2 TWBs, EU WFD, multi-national	5 TWBS, EU WFD, multi-national	2 TW cases, multi-national	1 TWB, EU WFD, multi-national	4 countries, EU	National basins, EU-WFD, 3 countries	9 case-studies, EU WFD, multi-national	INTL, focus – on 3 countries,	7 countries, EU WFD, multi-national	EU WFD, multi-national, focus – on 4 TWB
Legal base	Not applicable	Not applicable	Exists	Exists	Not	Not fully	Exists	Exists	R, CC, Intl	Exists	Exists	Mainly exists
Acting institutions**	N, R, CC, FI	N, R, CC, FI	N, R, CC, FI	R, CC, Intl	R, Intl	R	n/a	R, N, C, II		N,R, FI, CC, Intl	R,N,FI, RG	N, FI, R, Intl
Networking and cooperation***	2	2	2	1	2	1	2	2	3	3	3	3
Outputs/outcomes vs. regimes	policy advice	Policy advice	policy advice, scenarios	Models/tools, policy advise, assessment	Zoning, assessment, policy advise	policy advice, scenarios, assessment	policy advice, scenarios, assessment	PP tools and guidelines	PP guidelines, policy advise	MP for MPAs, formal base	Policy advise, tools and assessment	Policy advise, assessment, guidebook
Level of integration into policies****	2	2	2	2	2	2	3	1	3	3	2	3
Contribution to the existing/new TW regimes	Yes	Yes	Yes	Yes	Yes	Yes	Yes	Weak	Yes	Yes	Yes	Yes

Explanatory note to Table 2:

* - combination of section 2 (Manual) fields and domains: 1: Rivers, 2: Lakes, 3: Groundwater, 4: Coastal zone, 5: Wetlands, 6: Sea, 7: Wet & dry deposition, 8: Sediments

QN – quantity, QL – quality, E – ecology, M – morphology, S –sedimentation, EC – Economy, RM - Risk Management, LF - Legal Frameworks, GR - Governance Regime, PP – Public Participation

** - modification of section 1 (Manual): C – Citizens, N - NGOs, FI - Formal Institutions, II - Informal Institutions, CC - Companies & consultants, R – Research institutions, LG - Local government, RG - Regional government, NG - National government, WA – Water authorities, EU – European Union, Intl - International Institutions

*** - scores are given from 0 to 3, 0 – absence, 1 – weak, 2 – operational, 3 – excellent.

**** - scores are given from 0 to 3, 0 – absence (not at all), 1 – weak (some results are used but no real influence), 2 – medium (not all results are accepted/used), 3 – excellent (all major outcomes are integrated into policies).

Table 3: List of EU research and technological development projects supported by the EU RTD Programme and EU INTERREG/TACIS Programme

Name	URL	Subject index	Project Coordinator
EU FP4 Concerted Action Programme on the Effectiveness of International Environmental Agreements and EU Legislation (1999-2000)		EU environmental policy and legislation, regime effectiveness	Jørgen Wettestad, Senior Research Fellow, Fridtjof Nansen Institute
Research project on International Governmental Organisations and National Participation in Environmental Regimes: the Organisational Components in the Acidification Regime		EU environmental policy and legislation, regime effectiveness	
EU FP5 research project "Integrated strategies for the management of transboundary waters on the eastern European fringe - the pilot study of lake Peipsi/Chudskoe and its drainage basin" - MANTRA-EAST (2000 – 2003), Contract EVK1-CT-2000-00076	www.ctc.ee	Environmental Protection, Economic Aspects, Information Processing, Information Systems	Dr. Per Stalnacke, JORDFORSK – Centre for Soil and Environmental Research
EU FP5 Integrated water management of transboundary catchments - TRANSCAT	www.transcat-project.net	Other Energy Topics, Environmental Protection, Economic Aspects, Information Processing, Information Systems, Legislation, Regulations, Policies, Renewable Sources of Energy, Social Aspects	Marco Estrela, Instituto de Soldadura e Qualidade, Portugal
EU FP6 INCO project "Transboundary networks of marine protected areas for integrated conservation and sustainable development: biophysical, socio-economic and governance assessment in East Africa"	http://www.transmap.fc.ul.pt/	Education, Training, Resources of the Sea, Fisheries, Coordination, Cooperation, Evaluation, Legislation, Regulations, Policies	Fundacao da Universidade de Lisboa, E-mail: contact-this-project-via@cec.eu.int
EU FP6 INCO project "Concepts for integrated transboundary water management and sustainable socio-economic development in the cross border region of Albania, Former Yugoslav Republic of Macedonia (FYROM) and Greece" - TRABOREMA	http://www.traborema.net/	Education, Training, Coordination, Cooperation, Evaluation, Legislation, Regulations, Policies	MONTAN UNIVERSITAET LEOBEN, e-mail: contact-this-project-via@cec.eu.int
FP6 project "Institutional Mechanisms for Water Management in the Context of European Environmental Policies" – EUROWATER		Meteorology, Environmental Protection, Safety	Instituto Superior Tecnico
FP6 SSA "Empowerment and awareness building in river basin management through focus groups and citizens juries" - River Dialogue		Scientific Research, Information, Media, Social Aspects	Linköping University, e-mail: geogo@eki.liu.se
FP6 research project "Harmonising collaborative planning" – HARMONI-COP	http://www.harmonicop.info/	Environmental Protection, Economic Aspects, Social Aspects	University Osnabrück
10. FP6 Integrated project "Adaptation and Mitigation Strategies: supporting European Climate Strategy" – ADAM		Climate change, regime effectiveness	Tyndall Centre for Climate Change Research, UK
FP6 Integrated project "Reflexive Governance in the Public Interest" – REFGOV	http://www.adamproject.eu/		Centre for Philosophy of Law, (CPDR) of the Catholic University of Louvain (Louvain-La-Neuve)
Marine protected areas in the Eastern Baltic Sea (MPA in EBS)	http://www.bef.lv/?s=129&t=284	marine reserve, protected area, environmental training, environmental impact assessment, fishing industry	Coordinator: SO "Baltijas Vides forums", Latvia, e-mail: heidrun.fammler@bef.lv
BERNET - Integrated Management of Catchments in the Baltic Sea Basin.	www.bernet.org	water planning, classification, assessment, public participation, WFD implementation.	Fyn County, Oerbaekvej 100, 5220 Odense S.E., DENMARK , e-mail: OTJ@fyns-amt.dk

Helm and Sprinz demonstrated the empirical usefulness of their approach to measuring effectiveness of transboundary regimes that are aimed to reduce transboundary air pollution in Europe in 1980-90s. The international regime has been formed following signing the 1979 UN ECE Convention on Long-Range Transboundary Air Pollution (LPTAR), the 1985 Helsinki Protocol to the Convention and 1988 Sofia Protocol. The Helsinki Protocol required signatory countries to reduce their sulphur dioxine emissions or transboundary fluxes by 30 percent until 1993 relative to their 1980 emissions, the Sofia Protocol requires to control the nitrogen oxide emissions or their transboundary fluxes such as their 1994 values are no higher than those in 1987. To measure the regime effectiveness Helm and Sprinz used as the base and target years of the Helsinki Sulfur Protocol (1980 and 1993) and the Sofia protocol (1987 and 1994). Data on emissions, etc. were obtained through relevant international air pollution monitoring and research programmes. In addition to the data, expert interviews were conducted to assess emission reductions of the no-regime counterfactual. As a results of the calculations, the authors arrived to the aggregated effectiveness score 0.39 for the SO2 regime and 0,31 – for the NOx regime.

Helm and Sprinz concluded that the scores, although they are much higher than zero, are still lower than the theoretically assumed 100% (that would take place in the case if all signatory countries comply with taken responsibilities in the international regime). The overall conclusion of the exercise is that the LRTAP regime had a visible effect on the aggregate behaviour of countries for reducing the emissions with considerable variations across countries.

Concerted Action on the Effectiveness of International Environmental Agreements and EU Legislation

A scholarly discussion on the effectiveness of international environmental regimes was continued within the EU Concerted Action on the Effectiveness of International Environmental Agreements implemented in 1998 – 2000. The Action involved scholars of international relations who worked together in 1994 – 1997 in the project "*Implementation and Effectiveness of International Environmental Commitments*" - IEC and the above-mentioned research project on international governmental organisations and national participation in environmental regimes. In addition, European scientists in environmental policy analysis who have not been earlier involved in the network of regime scholars, took part in this Concerted Action. The Action therefore played an important role in promoting the regimes' research in Europe and further expanding the community of scholars in Europe involved in the international environmental regimes' (IER) research.

The work in the project was based on a classification of international environmental agreements, an inclusive definition of 'effectiveness,' and a preliminary list of factors to be considered. Multilateral environmental agreements were classified in three categories: (1) global environmental agreements (2) regional environmental agreements (involving significantly fewer than all countries), and (3) the environmental legislation of the European Community (i.e. involving a specific institutional structure to generate and implement the agreements). In an agenda setting paper, Konrad von Moltke (2000) formulated the following questions of the effectiveness research that demanded further attention of the scholars after the end of this Concerted Action:

(1) Identification of problem structure that the international environmental regimes address. Moltke identifies five distinct problem complexes where each of the complexes is addressed by a number of international regimes on different levels of governance. The five problem complexes are: nature conservation, marine pollution, global atmosphere, hazardous substances, and extractive resources;
(2) Identification of the institutions that are used to address the problem, and consideration of the fit between the two;
(3) Analysis of the legal and organisational issues that arise from this approach.

An important contribution of the Action to the development of the international regimes research was a study of interlinkages of environmental and economic international regimes, specifically of influences of the global markets and the economic globalisation to the international environmental governance, including formation and implementation of transboundary water regimes.

Research project on integrated strategies for the management of transboundary waters on the eastern European fringe - MANTRA-East project

An EU research project "Integrated Strategies for the Management of Transboundary Waters on the Eastern European fringe - the pilot study of Lake Peipsi and its drainage basin" - MANTRA-East was implemented in 2000 - 2003. The aim of the project was to analyse and develop strategic planning methodologies and scientific tools for the integrated water management in transboundary water basins located on the existing and future borders of the European Union in a context of the implementation of the EU Water Framework Directive. The project developed recommendations for institutional mechanisms and policy instruments for decision making on water management of transboundary watercourses and international lakes located on the fringes of the European Union.

Most of the studies of the MANTRA East project were conducted in the project pilot area – a transboundary water basin of the Lake Peipsi/Chudskoe[3] that is shared by Estonia, Russia and with a very little territory of the water basin also by Latvia and Belarus. The project has resulted in a thorough environmental assessment of the Lake Peipsi, an analysis of economic aspects with regard to water use, a study on the interest groups dealing with the use and protection of water; and integrated scenarios for the management of the transboundary waters in the Lake Peipsi basin. In addition, a GIS database has been created, comprising both Estonian and Russian parts of the Lake Peipsi region.

Within the project, a comprehensive overview of theories of international regimes and practices of the transboundary water cooperation was made by Gooch *et al.* (2002). A focus of the report is on the interaction of the environmental and human systems in transboundary water basins and especially on the actors involved in the cooperation. In the report chapter based on a description of the Spanish – Portuguese water management system, Gooch offers an analysis of the networks of actors involved in the transboundary cooperation, stressing an importance of values and beliefs of stakeholders involved in the transboundary water cooperation; and their perceptions of the environment and the transboundary water cooperation process.

Timmerman and Langaas (2005) organised studies of the information exchange and communication among the actors involved in implementation of transboundary water regimes within the MANTRA-East project. This included an innovative experimental study on the use and valuing of environmental information during the decision-making process by actors involved in transboundary water cooperation. The experimental study was conducted as a role-play of a theoretical decision making meeting of a transboundary water commission for the imaginary Lake Ispepi (≈ Lake Peipsi). Members of the Estonian – Russian transboundary water commissions and other stakeholders in the Estonian-Russian transboundary water cooperation were involved in the study. The meeting was asked to come to an agreement on the issue. As the level of trust between the two countries is rather low, the meeting could not come to a joint conclusion and it was argued that more information was needed to take a decision.

Studies of the role of stakeholders' networks in managing transboundary waters conducted within the MANTRA-East project included surveys of opinions of local inhabitants, focus groups, and in-depth personal interviews with experts in the transboundary Lake Peipsi Basin, on the both sides of the border – in Estonia and Russia. Based on those studies, mapping of major stakeholder groups, their interests and perceptions of water management issues in the transboundary water regions were conducted. The study emphasised an important role of networks of local stakeholders in the implementation of transboundary water regimes (see more in Roll 2003).

In the MANTRA-East project, a number of future scenarios of the environmental state in the Lake Peipsi basin have been constructed, to provide indications of how future developments on the Estonian-Russian border area may occur. The scenarios has taken into account a dynamics of the nutrient load into the lake, the lake retention and the resulting overall nutrient loads into the lake; the current and possible future changes in the economic and political situation in Estonia and Russia; and the regional development patterns. The scenarios were built upon the knowledge and results of both natural and social science approaches and using both qualitative and quantitative methodologies.

An important contribution of the MANTRA-East project to the studies of transboundary water regimes is in the studies of the actors involved in the transboundary water cooperation, the actors' beliefs, perceptions as well as different aspects of information exchange and communication among the actors. This shows that the cognitivist school was dominant in the project. The MANTRA-East innovative scenario approach that allowed integrating natural science and social science approaches, quantitative and qualitative information, is another important product of this project and an interesting approach that could be used to conduct interdisciplinary studies of transboundary water regimes in other water basin than the Lake Peipsi basin.

NeWater

The NeWater project "*New Approaches to Adaptive Water Management under Uncertainty*" contributes to the studies of transboundary water regimes in introducing a concept of the adaptive management (AM) of transboundary water basins. AM is in this project considered to aim at active learning and continually improving management strategies. Active learning includes gathering comprehensive knowledge of the current system and expected changes, e.g. by experimentation or simulation. Since relying only on the current knowledge base is not sufficient for taking water management decisions affecting the future, water management strategies need to be adaptable to new information and changing circumstances. The case study basins of the NeWater project are the Amu Darya, Elbe, Guadiana, Nile, Orange, Rhine and Tisza basin. Each basin is confronted with specific water management issues. In the Amu Darya, Orange, Guadiana and Nile river basin the issues are mainly related to water scarcity, whereas in the Elbe, Rhine and Tisza basin, pollution and floods are the central issues. Effective and efficient management of these issues requires

[3] Peipsi is the Estonian name for the lake and Chudskoe is the Russian name. For convenience, only „Peipsi" is used as a name of the lake further in the text.

transboundary cooperation and an appropriate institutional framework. A specific focus of the project is the linkage of transboundary waters and adaptive management.

The transboundary regimes of the basins under study are characterised by many similarities and differences. The most obvious similarity is that in every basin some form of structural transboundary cooperation (e.g. a river basin commission) has been established; the International Commissions for the Protection of the Rhine, Elbe and Danube River (ICPR, ICPE and ICPDR), the Orange-Senqu River Basin Commission (ORASECOM), the International Nile Commission (Nile-COM), the Interstate Commission for Water Coordination in the Aral Sea basin (ICWC) and the Commission for the Implementation and Development of the Albufeira Convention between Spain and Portugal (CADC). The Tisza and Amu Darya are sub-basins of the Danube and Aral Sea basin and the Guadiana is only one of the border rivers under attention of the CADC.

The NeWater project addressed the issue of evaluation of adaptive management in these transboundary water basins. To this end, an evaluation framework was developed that includes the formal and informal actor networks and their interactions, the legal framework that should include arrangements that support adaptive management, policy development and implementation that accounts for future developments, information management as an iterative process of determining information needs and collecting, interpreting, exchanging and using information, and an appropriate financing system that secures sufficient resources from private as well as public sources (Raadgever *et al.* 2007).

As the project preliminary results show, the tasks and responsibilities of the organisations differ strongly, as well as their functioning and effectiveness in reality. In the Rhine, Elbe and Guadiana basin, national governments have been the main initiating and financing parties. In the Amu Darya, Orange, Nile, and to a lesser extent also the Tisza basin, international donors have played a large role in the initiation and financing of basin organisations. It occurs that in the latter basins national governments are less committed and it is harder to develop and implement joint management strategies. Current transboundary cooperation in the African and Asian basins is therefore mainly aimed at developing trust and developing technical and institutional capacity. The role of informal actors in transboundary water management is in general limited, due to both distrust by governmental actors and limited capacities of the stakeholders.

The NeWater project included studies of information management in transboundary water basins, for instance studies of information exchange provisions in agreements concerning transboundary water management. According to the studies conducted, in the framework of the ICPR, ICPE, ICPDR, Nile-COM and CADC several working or expert groups have been established for this purpose. The ICWC has its own Scientific Information Centre and in the Orange basin the SADC-HYCOS contributes to information exchange and management. In practice, however, the production and exchange of information between formal actors has not been well-established in the Amu Darya, Guadiana, Nile, Orange and Tisza basin. Dissemination of information to stakeholders and the public is in general even more limited. Finally, the utilisation of information in transboundary decision-making is in many basins very limited, partly because the information production and exchange are not yet in operation.

The extent to which the regimes and information management in the studied basins support adaptive management varies significantly. A framework to assess the adaptiveness of river basin management regimes was developed in the NeWater project that was applied to a number of international river basins. This allowed identifying well developed and less adaptive elements of transboundary regimes. From this study it was concluded that the Rhine regime currently offers the largest potential for adaptive management, followed by the Elbe, Tisza, Orange, Guadiana, Nile and Amu Darya respectively (Raadgever *et al.* 2006). The transition from a traditional to an adaptive management regime can take decades. Although a first step was made in the basins by developing institutions for transboundary cooperation, implementation of the intended institutional structures is still ongoing. As long as the political setting is not ready for a real transition, there will be little determination for the development of transboundary water laws and policies. From the analysis it can be hypothesised that cooperation across administrative boundaries and joint information production are often part of the early phases of the transition towards AM. Somewhere in the middle of the transition, an appropriate legal framework, financing system, and policies should be developed and implemented, and a broad communication including public participation should be established. Requirements for adaptive management that are still hardly existent in any of the studied basins are adaptable legislation, cross-sectoral cooperation, scientific interdisciplinarity, cooperation between administrative levels, critical reflection on uncertainties, clear assumptions and adaptive mental models as well as appropriate utilisation of information.

This discussion shows that in the NeWater project the regime is considered from the cognitivist school, with emphasis on knowledge and information, social networks, and a strong institutional setting. The evaluation framework to determine the adaptiveness of regimes provides a more detailed view on different aspects of regimes but is also an example of the wide definition of a regime that is applied in the project. The framework is suited for evaluation purposes, but does not support approaches towards improving the regimes, especially where regimes focus more on seeking gains.

Research project "Harmonising collaborative planning" – HARMONI-COP

The general contents and aims of this project are described in greater length in Chapter 2 and 3 of this book. Regarding its transboundary dimension, the FP6 HarmoniCOP project (2002 -2005) focused on the actors networks in transboundary water management regimes – it was aimed at increasing the understanding of participatory river basin management planning (RBMP) in Europe in a view of a need to implement public participation provisions of the European Water Framework Directive. HARMONI-COP defined RBMP as an integrated cross-sectoral planning and management of river basins if necessary across political and administrative borders. A focus of the research in the project was on the information and communication between stakeholders and the social learning. The project thus builds on a cognitivist view of regimes. The project produced a handbook on public participation in water management.

Other EU projects addressing transboundary water regimes

There were also other projects that addressed issues of management of transboundary waters; those projects contained studies of regimes or provided case studies information that could be of help to scholars of international regimes in the future. Those projects include, for instance, TRANSCAT, TRABOREMA, EUROWATER, RIVER DIALOGUE, TRANSMAP, and other projects (See the list of EU projects in Tables 1 and 2). Besides the EU RTD projects, a number of EU INTERREG projects, including TRABANT and BERNET, provided useful empirical information for the further studies of regimes as well. The mentioned projects covered topics of water planning, GIS, public participation, water quality assessment etc. in transboundary context.

4.3 FUTURE PERSPECTIVES OF STUDIES OF INTERNATIONAL REGIMES IN EUROPE

In this chapter we discussed that research on international regimes in Europe concentrates around two schools; the neoliberal school and the cognitivist school. The neoliberal school is largely concentrated in a few research institutes in Europe (formerly at IIASA, Austria; currently at the Potsdam Institute for Climate Research, Germany and the Fridtjof Nansen Institute, Norway) and has a strong conceptual basis. The cognitivist school on the other side is less conceptually developed as there is no concentration of research in a few research centres. The cognitivist school comprises scholars who promote the adaptive management approach in their research. This approach allows them to address decision-making in water management in a more flexible and communicative way than the neoliberal approach. Adaptive management is nevertheless based on systems analysis principles that allows to see a larger picture of the water management process, and takes into account the links between social and ecological systems in water basins as well as includes perceptions of stakeholders into the analysis and the decision making process. Therefore, it is concluded that the two schools of thought are in a way complementary and depending on the political and geographical context of a transboundary water basin, one or the other approach would work more effectively. For instance, in regions where there is little trust or where even conflicts exist between countries, the cognitivist approach of regimes, which pays virtually no attention to the formal rules of cooperation that are important in developing strategies for conflict resolution and reestablishment of trust, falls short and the neoliberal approach would provide a better basis for improving the implementation and effectiveness of regimes.

From this conclusion we draw the following needs and priority directions of the research studies of transboundary water regimes. First, studies of approaches to the evaluation of transboundary water regimes' performance and regime effectiveness need to be further developed in Europe. Such studies have not only scientific value but also direct practical implication for a better understanding of the process of implementation of the European water policy and its central instrument – the EU Water Framework Directive. The studies of the international regimes in Europe could help not only to better describe regimes but also elaborate policy recommendations for more effective and efficient implementation of the European policies aimed at sustainable management of water resources in Europe and worldwide. Second, studies of transboundary regimes need to address issues of regimes scale, fit and interaction – for instance, regimes dealing with implementation of the EU foreign and security policies, and economic development - at one hand and transboundary water management at the other. Based on the research results so far (especially from the NeWater-project) it seems that transboundary regime research will have to be based on different perceptions and needs to develop specific research approaches when working on transboundary situations where not very cooperative countries are involved.

Third, studies of the role of scientific information and ideas and their influence to the behaviour of actors involved in international regimes (the cognitivist perspective) require development of the common methodology and consolidation of efforts of European researchers who are involved in those studies as well. Those studies should be more clearly linked to the studies of regimes as institutions; social science methodologies should be used more widely.

How should these studies of the transboundary water management regimes be organised? Looking at reports and articles on transboundary water management, it is clear that concerning the scholars dealing with the cognitive perspective of regime studies, common methodologies and definitions of studies have not sufficiently been developed. This makes comparison of results of studies of transboundary water management regimes produced within different projects and by different researchers, very difficult, often impossible. But to work towards AM in Europe we need to have an overall picture of transboundary water management regimes for the whole territory of Europe and on its borders. Therefore, elaboration of a methodology for the studies of the actors networks, stakeholder learning, information and communication, is a priority direction. Based on the studies of methodologies, common definitions, concepts and protocols for the data collection on case studies of transboundary water management regimes should be developed. These concepts and methodologies need to build on both the experiences in the neoliberal school as well as the cognitivist school.

The neoliberal perspective of studies of international environmental regimes is well established methodologically in Europe. Thanks to the IEC and IDGEC programmes, the research field and an international community of scholars of international regimes studies in Europe have been established. For the future development of the regimes' research in Europe, it is very valuable that within the IDGEC Programme Science Plan (2005) has been compiled. The Science Plan supports long-term sustainability and high quality of the national and international research of international regimes. In this area of studies it is today a good time to develop empirical studies analysing for instance implementation of the EU water legislation in transboundary water basins, including studies of effectiveness of transboundary water regimes. For comparability reasons it is important to stick to well-developed comprehensive methodological basis as well as the definitions and data collection protocols accepted in global projects such as IDGEC.

4.4 REFERENCES

Andresen, S., Skodvin, S., Underdal, A., and Wettestad, J. (2000) *Science and Politics In International Environmental Regimes: Between Integrity And Involvement.* Manchester University Press, New York

Brahm, E. (2005) International Regimes. In Burgess, G. and Burgess, H. (Eds.) *Beyond Intractability.* Conflict Research Consortium, University of Colorado, Boulder. Posted: September 2005 <http://www.beyondintractability.org/essay/international_regimes/>.

Breitmeier, H., Young O.R., and Zurn, M. (2006) *Analyzing International Environmental Regimes From Case Study to Database.* MIT Press, 336 pp.

Chalecki, E.L., Gleick, P.H. and others (2001) *Technologies, Institutions, and Social Issues in Arms Control and Transboundary Water-Resources Agreements.* Environmental Change and Security Program Report 8. Woodrow Wilson Center, USA.

Chayes, A. and Chayes, A. (1991) Compliance Without Enforcement. State Behaviour Under Regulatory Treaties, *Negotiation Journal*, 7 (3): 311-330.

David, V.G., Raustiala, K., and Skolnikoff, E.B. (Eds.) (1998) *The Implementation and Effectiveness of International Environmental Commitments: Theory and Practice.* Cambridge: MIT Press

Dworak, T. and Interwies, E. (2004) Linking Transboundary River Management to Water Security: the Example of Flooding in Europe. *The Forum - Magazine of the Bellagio Forum for Sustainable Development*, 11th edition, June 2004

Elliott, L. (1998) *The Global Politics of the Environment.* Macmillan Press Ltd.

European Environment Agency (2001). *Global International Waters Assessment* (GIWA), EEA: Copenhagen. Available at http://www.unep.org/dewa/assessments/ecosystems/water/vitalwater/12.htm#13

ESRC Global Environmental Change Programme (2000) *Who Governs the Global Environment?* University of Sussex, Brighton, UK

Geldof, G. (1994) *Adaptief Waterbeheer.* Deventer, Tauw Civiel en Bouw.

Geldof, G. D. (1995). "Adaptive Water Management - Integrated Water Management on the Edge of Chaos". *Water Science and Technology* 32(1): 7-13

Giordano, M. and Wolf, A.T. (2003) "Sharing waters: Post-Rio international water management". *Natural Resources Forum*, 27 (2003), pages: 163-171.

Goldstein, J. and Keohane, R. (Eds.) (1993) *Ideas and foreign policy: beliefs, institutions, and political change.* Ithaca: Cornell University Press

Gooch, G., Höglund, P., Roll, G., Lopman, E., and Alekseeva, N. (2002) *Review of existing structures, models and practices for transboundary water management.* MANTRA-East project report. Linköping – Tartu, Estonia

Grieco, J. (1990) *Cooperation among nations: Europe, America, and non-tariff barriers to trade*, Cornell studies in political economy. Ithaca: Cornell University Press.

Grieco, J. (1993) Understanding the Problem of International Cooperation: The Limits of Neoliberal Institutionalism and the Future of Realist Theory. In Baldwin, D.A. (Ed.) *Neorealism and neoliberalism: the contemporary debate.* . New York: Columbia University Press

Haas, P.M. (1992) Epistemic Communities and International-Policy Coordination - Introduction. *International Organization* 46 (1):1-35

Hasenclever, A., Mayer, P., and Rittberger, V. (1997) *Theories of international regimes*, Cambridge studies in international relations. New York: Cambridge University Press.

Hasenclever, A., Mayer, P., and Rittberger, V. (2000) Integrating theories of international regimes. *Review of International Studies* 26 (1):3-33.

Helm, C. and Sprinz, D. (2000) Measuring the Effectiveness of International Environmental Regimes. *Journal of Conflict Resolution*, Vol. 44, No. 5, 630-652

Hovi, J., Sprinz, D.F. and Underdal, A. 2003. The Oslo-Potsdam Solution to Measuring Regime Effectiveness: Critique, Response, And Extensions. *Global Environmental Politics* 3 (3):74-96

Keohane, R. (1984) *After hegemony: cooperation and discord in the world political economy*. Princeton, N.J.: Princeton University Press

Kibaroglu, A. (2002) *Building a Regime for the Waters of the Euphrates-Tigris River Basin*. Kluwer Law International.

Krasner, S. (Ed.) (1983) International Regimes. Ithaca, New York: Cornell University Press

Lidskog, R. and Sunqvist, G. (2002) The Role of Science in Environmental Regimes: The Case of LRTAP. *European Journal of International Relations*, Vol. 8, No. 1, 77-101

Lidskog, R.; Sundqvist, G. (2004). On the right track? Technology, geology and society in Swedish nuclear waste management. *Journal of Risk Research*, Vol. 7, No. 2, March, 251-268.

Lindemann, S. (2006) *Water Regime Formation in Europe: A Research Framework with Lessons from the Rhine and Elbe River Basins*. Available at SSRN: http://ssrn.com/abstract=925653

Miles, E. L., Underdal, A., Andresen, S., Wettestad, J., Skjærseth, J.B., Carlin, E.M. (2001) *Environmental Regime Effectiveness: Confronting Theory with Evidence*. Cambridge, MA: MIT Press

Magnuszewski, P.; Sendzimir, J.; Kronenberg, J. (2005). Conceptual Modeling for Adaptive Environmental Assessment and Management in the Barycz Valley, Lower Silesia, Poland. *International Journal of Environmental Research and Public Health*, 2 (2): 194–203

Mohr, L. (1998) *Impact Analysis for Programme Evaluation*. Chicago, IL: Dorsey.

Moltke, K. von. (2000) Research on the Effectiveness of International Environmental Agreements: Lessons for Policy Makers? In *Proceedings of the Concerted Action Programme's Final Conference*, Barcelona 200. Available at http://www.fni.no/

Mostert, E. Integrated Water Resources Management in The Netherlands: How Concepts Function. *Journal of Contemporary Water Research & Education.* Issue 135, Pages 19-27, Universities Council on Water Resources, December 2006

Pagan, O. and Crase, L. (2004) *Does adaptive management deliver in the Australian water sector?* 48th annual conference of the Australian Agricultural and Resource Economics Society, Melbourne, Victoria.

Porter, G., Brown J.W., and Chasek P.S. (2000) *Global Environmental Politic*. Westview Press, USA

Raadgever, G.T. and Mostert, E. (2005) *Analysis of Transboundary regimes - Comparative assessment of transboundary regimes and adaptive management in the NeWater case study basins*. Deliverable 1.3.1. of the NeWater project. Delft University of Technology, Delft, The Netherlands. (Draft)

Raadgever G.T., Mostert, E., Kranz, N., Interwies, E., Timmerman, J.G., 2007 Adaptive management in transboundary river basins: development and application of a framework for regime analysis. *Ecology and society* (submitted)

Raadgever, G.T., Mostert, E. and van de Giesen, N.C. (2006) Measuring adaptive river basin management. In M. Colosimo and D. F. Potts (Eds.) *Proceedings of the AWRA 2006 Summer Specialty Conference 'Adaptive Management of Water Resources'* (Missoula, Montana, 2006).

Roll, G. 2003. Interactive Management of Transboundary Waters on the External European Union Border. *Water Science and Technology*, 49. IWA Publishing, London, UK. Pp. 179-186

Ruggie, J.G. (1975) International Responses to Technology - Concepts and Trends. *International Organization* 29 (3): 557-583

Sendzimir, J., Magnuszewski, P., Balogh, P. Vari, A. (2006) Anticipatory modeling of biocomplexity in the Tisza River basin: first steps to establish a participatory adaptive framework. *Environmental Modelling and Software* 599-609

Sprinz, D. (2000) Research on Effectiveness of International Regimes: A Review of the State of the Art. In *Final Conference Materials of the EU Concerted Action on Regime Effectiveness*, 9 – 12 September 2000, Barcelona, Spain.

Sprinz, D. and Helm, C. (1999) The Effect of Global Environmental Regimes: A Measurement Concept. *International Political Science Review* 20(4): 359 – 369.

Timmerman, J.G. and Langaas, S. (2005) Water information - what is it good for? On the use of information in transboundary water management. Regional Environmental Change http://www.springerlink.com/openurl.asp?genre=article&id=doi:10.1007/s10113-004-0087-6 (online article)

Transboundary Freshwater Disputes Database. URL: http://www.transboundarywaters.orst.edu

Underdal, A. (1992) The Concept of Regime Effectiveness. Cooperation and Conflict 27 (3): 228 – 229.

Underdal, A. and Hanf, K., Eds. (2000) International Environmental Agreements and Domestic Politics: The Case of Acid Rain. Aldershot: Ashgate.

UNECE (1992) Convention on the Protection and Use of Transboundary Water Courses and International Lakes. Helsinki, Finland.

Van der Brugge, R., Rotmans, J., and Loorbach, D. (2004) The transition in Dutch water management. Regional Environmental Change http://www.springerlink.com/openurl.asp?genre=article&id=doi:10.1007/s10113-004-0086-7 (online article)

Victor, D. et al., Ed. (1998) The Implementation and Effectiveness of International Environmental Commitments: Theory and Practice. Cambridge, USA: MIT Press.

Wolf A.T., Natharius J.A., and others (1999) "International River Basins of the World". International Journal of Water Resources Development, Vol. 15 No. 4, December 1999.

Wouters, P.K., forthcoming. The Work of the United Nations on Water. 3 Volumes of Selected Works. Kluwer Law International.

Wouters, P.K., Vinogradov, S., and others (2005) Sharing Transboundary Waters - An Integrated Assessment of Equitable Entitlement : The Legal Assessment Model. IHP-VI, Technical Documents in Hydrology, No. 74, UNESCO, Paris, France

Yoffe, S., Fiske, G., Giordano, Mark and Meredith, Larson, K., Stahl, K., and Wolf, A.T. (2004) "Geography of international water conflict and cooperation: Data sets and applications". Water Resources Research, VOL. 40, 2004

Young, O.R. (1989a) International Cooperation: Building Regimes for Natural Resources and the Environment. Ithaca and London, Cornell University Press

Young, O.R. (1989b) The Politics of International Regime Formation - Managing Natural-Resources and the Environment. Ibid. 43 (3):349-375

Young, O.R. (1996) Creating Regimes: Arctic Accords and International Governance (Cornell Studies in Security Affairs). Ithaca: Cornell University Press.

Young, O.R. (1997) Global Governance: Drawing Insights from the Environmental Experience. Cambridge: MIT Press.
Young, O.R. (2001) Evaluating the Effectiveness of International Environmental Regimes. 2001. Global Environmental Politics. 1(1): 99 – 121.
Young, O.R. (2002) The Institutional Dimensions of Environmental Change. Fit, Interplay, and Scale. Cambridge: MIT Press.
Young, O.R. (2005) Science Plan. Institutional Dimensions of Global Environmental Change. IHDP Report No. 16. Bonn, Germany, Revised edition: July 2005
Young, O.R. and Osherenko, G. (Eds.) (1993) Polar Politics – Creating International Environmental Regimes. Ithaca: Cornell University Press.
Young, O.R. (Ed.) (1999) The Effectiveness of International Environmental Regimes. Cambridge, MA: MIT Press, 1999.

5

Economics of integrated water management in a European perspective

Ing-Marie Gren

5.1 INTRODUCTION

Aquatic ecosystems provide society with a number of valuable services such as drinking water of sufficient quality, irrigation water for agriculture, recreational values, and biodiversity. Degradation of the aquatic systems implies losses in these services, and can be counteracted only by implementation of costly measures. However, since most aquatic ecosystems are common property resources, where benefits from measures are enjoyed by many users but the costs are born only by those implementing the measures, they are usually not managed for the best of the society. This has been recognised in environmental economics for decades, and the question has been raised on how we can implement costly measures for combating water degradation with an efficient use of societies' total resources. Today, a few agreements emerge from this literature. One purpose of this chapter is to investigate if and, if so, how these insights and methods have been developed and applied to integrated water management of European aquatic systems. Another purpose is to identify knowledge gaps and needs for future research directions on integrated water management within a European context.

However, as in all economics, there is a need for a stringent understanding of the environmental problem at hand in order to find appropriate analytical and empirical tools. In principle, from the introductory chapter in this book, integrated water management is characterised by many dispersed sources of different polluting substances, such as nitrogen and phosphorous, to the water recipients. This, in turn, is likely to lead to conflicting interest with respect to the achievement of water quality and quantity targets in groundwater, rivers, coastal waters and lakes. Furthermore, for transboundary aquatic systems such as the Baltic Sea, the Rhine and the Danube rivers, several sovereign states may claim rights to the use of ecosystem services provided by aquatic systems in the basin. The drainage basin scale and the dispersion of emission sources, which may affect several aquatic ecosystems, also implies risk and uncertainty with respect to predicting impacts on the systems from different management strategies, climatic change and other forces. What then can economics say about adaptive water management under stochastic and uncertain changes? Economics has contributed to the issue of adaptive management since the mid 1900s, but on other needs, such as identification of appropriate policies for adaptation of national economic activities to internationally driven changes in energy supply and on international financial markets. One of the main issues is then how to set the playground for all involved actors with conflicting interests so as to stimulate the best use of all society's resources in the short and long term periods. Thus, economics is not so much about telling individual actors what to do, for example which water cleaning measures to adopt, but does instead contribute to adaptive water management by suggesting institutional frameworks for managing conflicting needs and promoting efficient use of total resources under conditions of uncertainty. More precisely, economics of complex aquatic systems deals with two main issues: *i)* identification and calculation of cost and benefits from projects aimed at improving water conditions, and *ii)* analyses and comparison of different

 The adaptiveness of IWRM; Analysing European IWRM research Edited by Jos G. Timmerman, Claudia Pahl-Wostl and Jörn Möltgen. ISBN: 9781843391722. Published by IWA Publishing, London, UK.

institutions and polices. The first issue is much related to the question on the level of ambitions of management programs and the second concerns how programs should be implemented. Generally, a distinction is being made depending on the possibility of a public body able to implement and supervise compliance of policies. For transboundary water, there is no such super state power, and implementation of cleaning programs must be made on a voluntary basis through bargain among the riparian countries. Aquatic systems within national boundaries can be implemented by public authorities. Economic analysis of all these issues deal with risk and uncertainty and with spatial and dynamic spread in the location and implementation of policies and specific measures, but the empirical application to European waters differ. Based on these issues, this book chapter is divided into four main sections:

- ambition level of management programs, which involves questions on how much resources to spend on water management,
- choice of policy instruments, or how to choose the best policy design for adaptive management,
- transboundary waters, which includes issues on how agreements on management of international common water resources can be made and sustained, and
- risk and uncertainty, which is at the core of adaptive management, treat issues on how to account for uncertainty when setting targets and choosing policy instruments.

Each chapter gives a brief economic theoretical background and then presents examples from applied research made for European aquatic systems with a focus on EU funded projects. Although the theoretical illustrations presented for each of the four research fields may seem unrealistically simple they are of practical use since they give guidance to choice of strategies and data needs. The chapter will not provide an exhaustive list of all relevant research. Instead only examples representing typical fields of research are highlighted. The choice of EU funded projects is based on records in the EU Cordis data basis by applying a number of different relevant key words, such as 'policies', 'benefits', 'water management', but always including 'economics' since the purpose of the chapter is to present economics of water management, see appendix for a list of included EU projects and their acronyms. This chapter ends with a concluding discussion on knowledge gaps and associated economics research needs for the management of European waters.

5.2 CHOOSING AMBITION LEVELS AND FUNDING NEEDS

A traditional economic approach to identifying the 'right' ambition level and associated financial need is to find that level of, say, cleaning of a lake, which maximises social total net benefit. This does not necessarily mean that all possible cleaning should be made, but depends on the change in benefits and costs associated with higher ambition levels, which is illustrated in Figure 1.

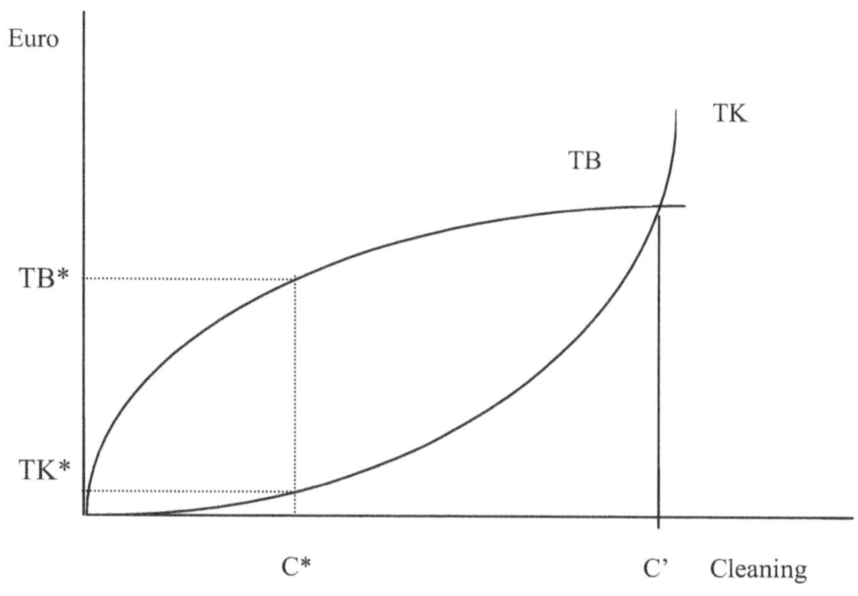

Figure 1: Illustration of optimal cleaning level and associated maximum net benefits and funding needs

The curve *TB* illustrates total benefits from increased cleaning levels, which should include values of all ecosystem services. The curve illustrates the common feature of declining benefits, which implies that additional benefits from improved cleaning become smaller with increasing ambition level. The *TK* curve shows the allocation of different cleaning measures such as creation of wetlands, buffer strips, and reduction in pollutant loads, which minimises total costs for each cleaning level in the lake (this is discussed more below). Equivalently, the curve shows the minimum financial resources needed for achieving each cleaning level. The maximum net benefits are obtained at the largest difference between *TB* and *TK*, which occurs at *C**, where max net benefits is *TB**-*TK**. The minimum financial resources needed for achieving this level are *TK**, which can be paid by public bodies by means of compensation payments to firms bearing the cost of the measures and/or by the firms undertaking the cleaning measures. Note that all cleaning levels up to *C'* in Figure 1 give positive net benefit, but only *C** gives the highest net benefit.

Thus, equipped with benefit and cost functions of specific programs, which can include several aquatic systems in a drainage basin, the choice of ambition level seems rather straightforward. However, in practice it is far from a trivial issue to quantify these functions, since this requires quantification of links between changes in land use and pollutant emissions in the drainage basin with water quantity and quality impacts.

Starting with research on cost of programs, an old economic wisdom is that abatement measures should be allocated among these sources where their marginal abatement costs are equal in order to achieve stipulated targets at minimum costs, so called cost effectiveness analysis. Such cost effective solutions underlie the hypothetical *TK* curve in Figure 1. If the condition of equal marginal costs is not fulfilled, reallocation can be made by increasing abatement at measures with relatively low marginal abatement costs to measures and decrease by the same amount at sources with relatively high marginal abatement cost. Total abatement is then unchanged, but is achieved at a lower total cost. It is well known that such an allocation can be implemented by the means of economic instruments such as charges, subsidies and market for emission rights, see discussion in Section 5.3.

Cleaning cost of an aquatic system is determined by the abatement cost at the source and its impact on the sea. For example, consider abatement at two sewage treatment plants, one located close to the water recipient and another situated up-streams in a drainage basin. Cleaning costs at the sources are the same for both these plants. However, in order to achieve a given abatement level at the sea, less cleaning is needed at the coastal than at the up-streams plant. The reason is that a part of the upstream nutrient discharges into a stream would have been cleaned by the denitrification, sedimentation and plant assimilation processes before reaching the Sea. This is not the case for a plant located at the coast. Then, no cleaning by nature is available free of charge. The impact on the Sea from this source from a unit anthropogenic cleaning is then higher than for the up-streams source. This means, in turn, that marginal abatement cost is lower for the coastal plant than for the up-streams located plant.

The EU funded Baltic Drainage Basin Project (BDBP) is the first study of large-scale cost effective water quality management by reduction in nutrients. The Baltic drainage basin was divided into 23 different drainage basins, where measures involved agriculture, sewage treatment plants, construction of wetlands, air emission from mobile and stationary combustion sources. In total, 15 measures from each of the 23 drainage basins were included, for each of which costs at the emission source and impact on the Baltic Sea were calculated. Based on this a programming model was constructed to calculate cost effective solutions for different requirements on reductions in per cent of nitrogen and phosphorous, see Figure 2.

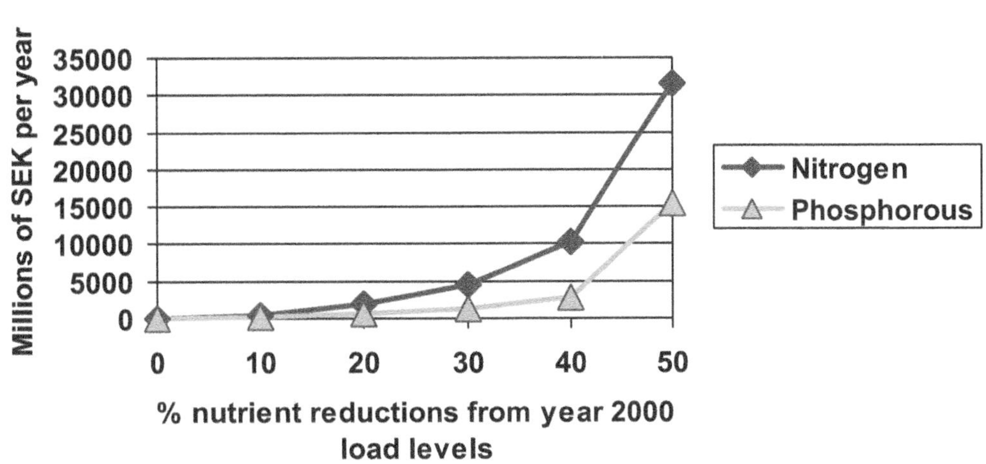

Figure 2: Minimum costs for different reductions in nitrogen and phosphorous loads to the Baltic Sea. (Sources: data from Gren *et al.* 1997, Gren and Folmer 2003)

The results presented in Figure 2 indicate that phosphorous reductions are considerably less costly than the same percentage reductions in nitrogen. The main reason for this difference in costs is that a relatively large share of phosphorous loads originates from point sources with direct impact on the sea.

Several subsequent EU funded projects calculate costs of specific measures, such as costs of improved technologies for waste water treatment (AGROIWATECH), costs of provision of drinking water in Cyprus (Swanson(b)), and for improved water quality system (SWIFT-WFD). There are very few studies calculating and comparing costs of several different cleaning measures, such as subprojects within the ERMITE and CLIME projects.

However, whether or not measures should be implemented is based, not only on their costs, but also on associated social values, i.e. the *TB* curve in Figure 1. Some of these values can be traded on the market, such as drinking water in bottles. Since several of the values obtained from water quality improvement, such as improved bathing opportunities and biodiversity, cannot be traded, their assessment needs to be obtained in other ways. There is a large literature on the valuation of environmental changes, which has developed rapidly during five decades (see Turner *et al.* 2003 for a review). In principle, we identify two types of approaches: preferences obtained from hypothetical markets and preferences obtained indirectly through real markets. Both these methods have their pros and cons, but they share two difficulties. One is how to find measurement on a continuum of environmental changes, and not only one or few, and the other is how to treat stochastic and uncertain environmental changes for preference formation. Methods have been developed in order to solve this problem, but the problem of measuring preference formation under stochastic and long term conditions remain.

In the applied economics research on aquatic systems in Europe as well as in other parts of the world, non-market values have to a large extent been estimated for wetlands and coastal zones, where recreational values have been calculated for both ecosystems and pollutant cleaning values for wetlands. A specific feature of studies made on European waters are the large scale benefit cost studies of the Baltic Sea, which emerged from the EU funded Baltic Drainage Basin Project (BDBP), see Table 1.

Table 1 presents some valuation studies of European aquatic systems, which are obtained from the selected EU projects.

Table 1: Examples of valuation studies of European aquatic systems.

Study **(name of EU project's principal investigator or acronym in parenthesis)**	***Region and type of valuation***
Söderqvist 2000 (BDBP)	Baltic Sea, total value from reduced eutrophication. Valuation study made in Sweden
(Pearce) *No reference included! It is in the list of EU projects*	Benefit transfers of water quality health values among European countries
Markowska *et al.* 2000 (BDBP)	Baltic Sea, total value from reduced eutrophication, study made in Poland
Söderqvist *et al.* 2000 (ECOWET)	Meta analysis of wetland values
Byström *et al.* 2000 (ECOWET)	Insurance value of wetlands for stochastic nitrogen pollution in Sweden.
Bateman *et al.* 2005 (CLIME, and EMERGE)	UK and Scotland, values from reduction in acidity of lakes
Bateman *et al.* 2006 (CLIME)	England and Wales, values from reduced eutrophication in surface waters
Gren *et al.* 2007a (EVALUWET)	Multifunctional values of wetlands in a landscape context, Sweden

Starting in the 1970s, values of wetlands have been estimated during a relatively long period. The EU funded ECOWET project was pioneering by its accomplishment of empirical meta analyses in the 1990s, see Table 1. Since then, several meta analyses have been carried out. Bateman *et al.* 2005 in Table 1 provide one of the few attempts to estimate, not only total value of a single environmental change, but also marginal values for different changes in, in this case the number of clean lakes in UK and Scotland, which allows for the derivation of benefit cost functions illustrated in Figure 1.

Söderqvist (2000) and Markowska *et al.* (2000) provide early large-scale valuation studies of reductions in eutrophication, which is followed up by Bateman *et al.* (2006) but for other water systems. The benefit transfer project carried out by the principal investigator David Pearce was also a pioneering EU funded project, which preceded a number of benefit transfer studies. Another unique EU study is Gren *et al.* (2007a), which, unlike most studies on water quality and wetlands, includes valuation of wetlands in a landscape context with different conflicting uses of environmental services such as water quality and livestock production. Byström *et al.* (2000) is among the first studies estimating the potential value of wetlands as buffering against stochastic water pollution. The recently initiated

AQUAMONEY project is also innovative by its attempts to evaluate and assess monetary benefits to multifunctional uses of water in a WFD context.

This brief presentation of empirical studies of costs and benefits from improved European waters reveals that cost functions as illustrated in Figure 1 have been calculated for different large and small-scale water. However, associated estimated benefits functions do not exist, which implies that it is in practice impossible to identify the ambition level which gives the highest return per unit of investment in water management. Instead, the valuation studies allow for point estimates, i.e. benefits and costs for a particular cleaning level that tells only whether or not a specific water management program passes the benefit cost test. Referring to Figure 1 this means that all cleaning levels up to *C'* would be accepted.

5.3 CHOOSING THE BEST WATER POLICY INSTRUMENT

Starting in early 1930s, the literature on how to design environmental policies has developed rapidly. One prerequisite, in addition to the understanding of the water issue as such and the definition of target(s), is to identify which criteria the policy instruments are supposed to achieve. Common criteria used for policy analyses are

- cost effectiveness and technological development, which means that the instruments should generate determined target(s) at minimum cost in the short run and in the long run by stimulating technological development,
- equity and fairness, implying that agreements should result in equal relative cost burdens among the partners and adopt the Polluter Pay Principle (PPP),
- precision and flexibility, the instrument should result in actions as close to the target as possible, but at the same time be flexible enough to give the opportunity of changing targets if found necessary in an environmental perspective, i.e. adaptive management.
- low transaction costs, i.e. costs for administration, implementation and supervision of policies

Is there any instrument that satisfies all these criteria? Unfortunately not, but there are candidates that perform relatively well with respect to several of the criteria. The choice is then in principle between three types of policies – command and control, economic instruments, and markets for nutrient trading –, which are briefly discussed in the following.

The command and control policy, which has been much used in practice, implies that a regulator decides which measures, such as technology standards like the best available technology, or bans on uses of certain chemicals, to implement. This is the type of policy used for water quality management in most European countries. It is usually the most costly policy since it gives no room for cost adjustment at the firm level. On the other hand, given that all firms comply with the regulation, the precision is high.

Economic instruments give incentives to either reduce polluting activities, such as fertiliser charges, or make it beneficial to adopt cleaning activities, such as subsidies for catch crops. Such instruments rely on the PPP principle, and are also cost effective since firms have the possibility to adjust their activities according to their control costs. However, the precision is relatively low since it is difficult to predict the final outcome on nutrient loading from firms' adjustments to the instrument. On the other hand, there is a great degree of flexibility since it is in principle relatively easy to change the charge level, increasing it if the final loads are too high and vice versa.

The third option – nutrient trading market – is a combination of command and control policy and economic instruments where maximum pollutant allowances are set for a specific region, pollution quotas are then distributed to all affected firms, but the difference to a command and control policy is that these quotas, or permits, can be traded between the firms. This means that firms with high costs purchase permits and low costs firms sell if the offered price exceeds the corresponding abatement cost. Thus, cost effectiveness is achieved through the allocation of abatement towards low cost firms. In principle then, the nutrient trading market has all the relative advantages attributed to both the command and control and economic instrument systems.

Since the achievement of cost effectiveness has been a central argument from economists in favour of economic instruments and permit market systems, its functioning is illustrated in Figure 3.

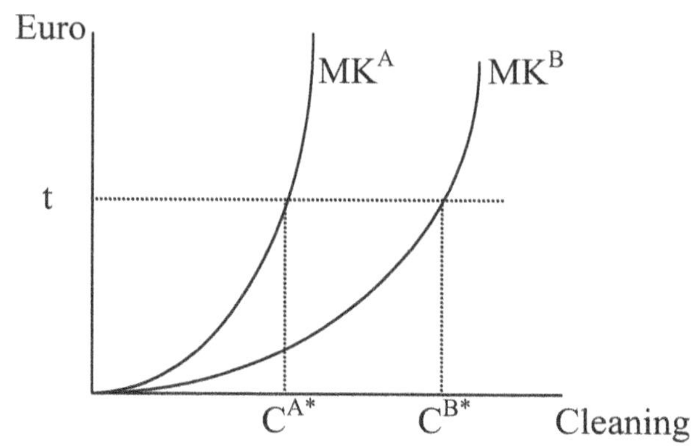

Figure 3: Cleaning responses of two firms, A och B, with marginal costs for cleaning MK^A and MK^B, at the charge/permit price of *t*.

The two curves, MK^A and MK^B, illustrate how cleaning costs increase for firm A and B respectively when cleaning increases by one unit. Common to both firms is that as marginal cleaning costs increase the higher is the achieved cleaning level, which is a common phenomenon in practice with for example sewage treatment plants. An increase in nutrient cleaning from 0 to 10 per cent of incoming nutrients is less costly than an increase from 60 to 70 per cent. Assume now that the charge per unit of pollution, *t*, is introduced. Both firms then respond by cleaning at the level where $t=MK^A$ and $t=MK^B$, which occur at C^{A*} and C^{B*} respectively. The reason is that each firm clean up as long as it is less costly than paying the charge, which hold for all cleaning levels below C^{A*} and C^{B*}. At higher cleaning levels it is less costly to pay the charge since then $MKA > t$ and $MKB>t$. Since such responses from all firms imply that they clean where their marginal costs are equal, at $MK^A=MK^B$ in Figure 3, the condition for a cost effective solution is fulfilled. It is then not possible to reallocate cleaning between the firms without increasing the total cleaning cost. Another argument in favour of economic instruments is cost effectiveness in the long run. Since it is costly to pollute, incentives exist for finding new pollution saving technologies.

However, when measures and sources have different impacts on the recipient, which is typical in a IWRM context due to the dispersed location of sources, it is less obvious which instrument is best. The reason is that in order to achieve cost effectiveness with heterogeneous sources, charges or permit prices must be designed according to each source impact on the water recipient. This heterogeneity among sources has concerned economists since the 1960s. The problem is that cost effective allocation of abatement measures then requires policy instruments, such as an effluent charge, that is unequal among sources due to their differences in impacts on the water recipient. Why then should this be a problem? It is in principle possible to design a policy system where charges differ among, for example farmers or sewage treatment plants. The problem arises from the need of monitoring all sources, which can result in relatively large transaction costs. There is thus a trade off between low abatement costs and high transaction costs. Although this is well known in theory, there are very few studies which estimate transaction costs under different policy design. Instead there is a large empirical literature on estimation of costs from uniform policies, which are calculated as the difference between cost under a uniform system, such as uniform proportional reduction requirements among regions or sectors, and cost effective solutions by means of mathematical programming models. For large differences, a uniform system is not justified. Such calculations of some drainage basins in Europe reveal large variations in relative efficiency losses, see Table 2.

Table 2: Differences in total costs between uniform and differentiated instruments as per cent of total costs in a cost effective solution.

***Study* Author (EU project acronym)**	***Water area***	***Uniform charge or permit price***	***Uniform cleaning requirement***
Gren 1993	Drainage basin of Stockholm archipelago		233
Gren *et al.* 1997 (BDBP)	Baltic Sea		40-110
Brännlund and Gren 1999	Coastal zones of Sweden	10-15	
Elofsson 1999	Baltic Sea		75
Brady 2003	Coastal zones of southern Sweden	19	
Scharin 2005a	Drainage basin of Stockholm archipelago	56	5 – 61
Scharin 2005b	Drainage basin of Stockholm archipelago	10-30	25-120

All calculations presented in Table 2 are made for waters with damages from eutrophicaion and include abatement measures for the agricultural sector. In addition to spatial scale, they differ with respect to choice of abatement measures in other sectors than agriculture and inclusion of pollutants. Brännlund and Gren (1999) and Scharin (2005a, b) calculate differences in costs between uniform and differentiated charges on nitrogen fertilisers in different Swedish counties. The studies with the largest spatial scale of the drainage basins are Gren *et al.* (1997) and Elofsson (1999), where costs of uniform and differentiated nitrogen reductions are calculated and compared. One difference between the studies is variation in measures. Whereas Elofsson (1999) includes agriculture and sewage treatment plants, Gren *et al.* (1997) also account for stationary and mobile combustion sources, construction of wetlands, and water discharges from industry. Another is that Gren *et al.* (1997) includes both nitrogen and phosphorous reductions while Elofsson (1997) contains only nitrogen. The Gren *et al.* (1997) study is the only one resulting from a EU funded project.

The range of per cent increases for single studies is due to calculations of costs for different reduction levels. In general, the relative costs of uniform measures increase up to a certain level, but then decline as the requirements become so stringent that all measures have to be implemented for reaching the target. Gren *et al.* (1997) and Elofsson (1999) calculate cost differences for nitrogen and phosphorous reductions of 50 per cent. The results presented in Table 2 indicate that uniform reduction requirements can be significantly more expensive than a uniform charge. The reason is that, unlike a uniform reduction requirement, a uniform charge allows for the adjustment of costs at the firm level as illustrated in Figure 2. Thus, considering transaction costs, the results in Table 2 indicate that efficiency losses of a uniform charge system among emission sources are likely not to compensate for transaction costs of differentiated charges. On the other hand, the relatively large efficiency losses of a command and control system with requirements of equal proportional cleaning requirements are likely to exceed the transaction costs of a differentiated system.

When comparing policy studies with the number of EU funded valuation studies, there are much less EU funded projects with economics focus on institutions and policy for integrated water management. The BDBP in Table 2 is an early large-scale study covering several countries. Other relatively early projects are Swanson (b), where different policies are compared for provision of drinking water in Cyprus, and Swanson(a) which identifies policies against toxic waste. Related projects are those identifying efficient policies for air and water pollution or natural resource use from specific sectors, where IMPRINT-NET is an example of a project where policies for emissions from the transport sectors are investigated and the project by Post, WADI and WATERWEB identify impacts on farmers' incomes from water management in different European drainage basins. On the other hand, there are relatively many projects aimed at linking different activities, most often different land uses, with impacts on water quality and quantity. For example, the CHESS project builds models of nutrient transport in several drainage basins and then simulates impacts on different land use activities and waters from alternative climate change scenarios. The GOUVERNE project also develops drainage basin models, which are designed for use as decision support tools for stakeholders. Design of decision support tools including physical and biogeochemical data is also at focus for the COAST project applied on coastal management. However, none of these related projects applies a holistic approach, including several regions and production sectors, where different institutional designs are evaluated and compared with respect to different criteria of comparison, such as cost effectiveness and/or technological development.

5.4 TRANSBOUNDARY WATER MANAGEMENT

Many rivers, lakes, and seas share their drainage basin by sovereign countries. The associated interdependency among countries can be categorised in two types: unidirectional and mutual linkages (Mäler 1990). Water quality and quantity in rivers, such as Rhine and Danube, provide examples of unidirectional linkages between upstream countries, whose water management affects other, downstream, countries. Such linkage implies that water quality improvements made by the upstream country are beneficial also for downstream located countries. If the country considering management actions does not include the benefits of other countries benefits are likely to be too small to cover the management costs. This is illustrated in Figure 4 .

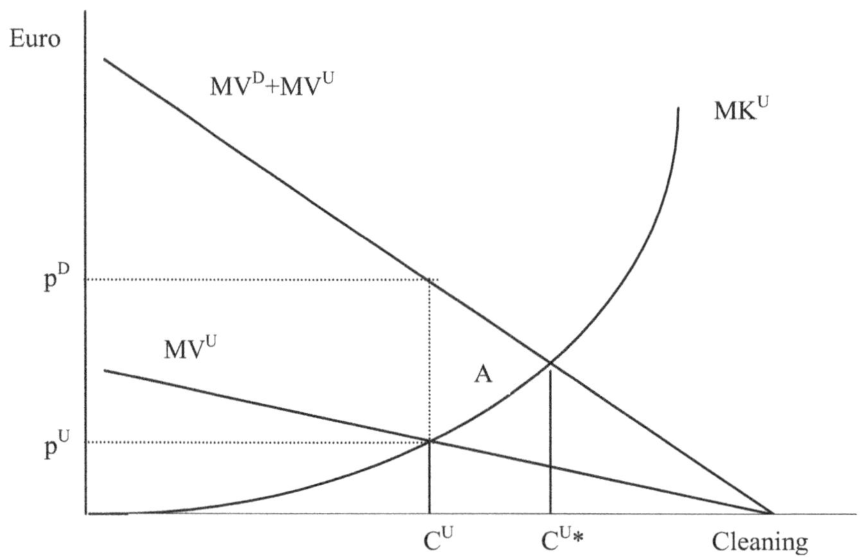

Figure 4: Illustration of upstream water cleaning

Similar to Figure 1, the MK^U curve illustrates the increasing marginal cost of cleaning for the upstream country *U*. The MV^U curve shows the marginal values from water quality improvement of the upstream country, the shape illustrates the declining marginal values. That is, the water quality improvement of, say, the first tons of nutrient reductions are higher than at higher level of reductions. However, since also the downstream country benefits from upstream water quality improvement, MV^{D}, the total marginal value of cleaning is the sum of the values of water quality improvements in the downstream and upstream countries, illustrated by the curve MV^U+MV^D. If the upstream country considers only its own costs and values from cleaning, it will end up cleaning at the level C^U. This is because cleaning is undertaken as long as the additional value, i.e. MV^U, exceeds the marginal cost MK^U. At levels exceeding C^U additional costs are larger than associated values. However, in a drainage basin context further values can be obtained by more cleaning due to the benefits obtained by the downstream country, which is illustrated by the dashed line connecting MK^U and MV^D+MV^U at C^U, showing the marginal value for the downstream country. Thus, by increasing cleaning to C^{U*} the net value would, in a drainage basin context, increase by the triangular area denoted by *A* in Figure 4.

Would it then be possible to obtain additional value from cleaning of an upstream country obtained by the other downstream country? According to the Coase theorem, this is made possible by construction of well defined property rights, by means of which the two countries could bargain on further cleaning (e.g. Perman *et al.* 2002). The theorem states that the reason for an agreement is that the upstream or downstream country will accept a bid from the other country, which covers its marginal cost or marginal value of cleaning. At C^U in Figure 4, the downstream country is prepared to pay up to p^D for additional cleaning by one unit. The upstream country accepts this since the bid covers the marginal cleaning cost of p^U. Note that the upstream country would accept a lower bid than p^D as long as it exceeds p^U. The final bid is determined by the bargaining powers of the two countries. If these are equal, the bargaining process will end up in cleaning at C^{U*}, which maximises total net values at the drainage basin scale. Thus, potentially significant gains could be made from the establishment of institutions that facilitate participation of stakeholders in the bargaining processes. Participation of stakeholder is discussed in-depth in other chapters in this book

When there are mutual linkages between countries some or all countries affect and are affected by their own and others water management actions. Examples of such dependencies are the Baltic Sea and the North Sea. Then the cleaning of one country, say Poland, affects all other riparian countries, and Poland is in turn affected by cleaning in the other countries due to the transport of nutrients in the sea. What will then be the final outcome of cleaning for the sovereign countries in such a setting? In order to illustrate this, assume only two countries, Poland and Finland, which have the choice of saying Yes or No to a common cleaning project. The outcome of each country depends on the other country's choice, which is illustrated in a hypothetical situation in Table 3.

Table 3: Illustration of outcome under a one-shot game between two countries

		Finland: Yes	No
Poland:	Yes	10 ; 15	- 3 ; 17
	No	11 ; -3	0 ; 0

The number in each cell shows the payoffs to each of the countries under different actions from the other country. The first number in each cell illustrates Poland's net pay off and the other Finland's. Thus, the pair of number 10;15 in the cell where each country says Yes to the cleaning program shows that Poland receives 10 and Finland 15. The higher number for Finland is a reflection of the direction of the stream in the Baltic Sea. Cleaning in Poland has a higher impact on Finland than vice versa due to the northeast directions of the streams and pollutants. It is assumed that the net pay offs are zero for the countries if they both say No to the cleaning program.

Table 3 illustrates a typical free riding option for both countries. Consider the case when Poland says No and Finland Yes. The net gains to Poland are 11, which are higher then if both say Yes to the cleaning program. Poland thus makes some gains from avoiding the costs of the cleaning but can make gains from the Finnish cleaning. Similarly, Finland also has a free riding option, since the gain from saying No when Poland says Yes, i.e. 17, is higher than when both say Yes. If the countries have a situation where they can choose only one time, they will end up saying No to the cleaning program, although the total gains are highest if both say Yes. However, under other institutional settings when partners can renegotiate, the outcome with the largest net benefits is likely to be final solution (e.g. Mäler 1990). Again, we notice the need of institutional settings for participation and bargaining of different stakeholders.

What then can we say about the differences in bargaining outcomes in empirical settings? Although there are relatively many studies calculating net benefits for different cooperative and non-cooperative outcomes for transboundary air pollution and climate change problems (see Barrett (2005) and Kolstad and Toman (2005) for surveys of theory and empirical applications), there are few such studies on transboundary waters. Most of these are applied on the Baltic Sea (Gren *et al.* 1997, 2003; Elofsson 2003) and one on the Black Sea (Bayramoglu 2006). All Baltic Sea studies emerged from the EU Baltic Drainage Basin Project (BDBP), and they all show that Poland would make net gains from not participating in international cooperations on nitrogen and phosphorous reductions because cleaning costs are relatively low and the share of unregulated loads of each of the two nutrients is relatively large. These results remain when stochastic load is considered, which is discussed in Section 5.5. Furthermore, total costs from reducing nitrogen and phosphorous load by 50 per cent could decrease by 60 or 25 per cent for nitrogen and phosphorous reductions respectively if all countries cooperate as compared to national actions (Gren *et al.* 1997). Bayramoglu (2006) calculated outcomes from different institutional settings for two countries – Romania and Ukraine – in the Black Sea drainage basin. An interesting dynamic approach is applied where changes in nitrogen load are linked to fishery reproduction and harvesting. Each country then considers costly actions for reducing nitrogen loads which affect the benefits from fish harvesting of both countries. The results indicate that Romania makes the largest gains from deviating from the cooperative solution.

This brief review of studies of transboundary water management indicates that relatively few studies with the purpose of calculating net benefits for different actors have been applied to European waters and carried out within the framework of EU funded projects. TRABOREMA is a EU-project dealing with a transboundary common water system shared by Albania and Greece, but is focused on the design of monitoring systems. The relatively low number of studies on economics of transboundary water management contrasts the valuation studies, which are common within EU-projects on water management, but is in accordance with the small number of projects on policy design.

5.5 MANAGEMENT UNDER RISK AND UNCERTAINTY

As pointed out in several chapters of this book, predictions of impacts of management strategies, climatic change etc. on water quality can be made only under conditions of risk and uncertainty. In economics, risk is associated with the ability to assign probabilities to different outcomes whereas uncertainty refers to a situation with incomplete information where neither all outcomes nor probabilities to known outcomes can be inferred (e.g. Knight 1921). People can then show risk and uncertainty aversion, or being neutral to both. In rare cases, risk or uncertainty loving behaviour occurs, such as different kinds of gambling. In the more common situations of risk averse attitudes, people prefer a relatively certain outcome over outcomes associated with risk. Due to this risk aversion, they are willing to pay a premium for decreasing or avoiding the risk, the size of which depends on the degree of risk aversion. Uncertainty aversion can be traced to Ellsberg's paradox (Ellsberg 1961). The paradox revealed that people, in an expected utility maximising framework, make seemingly irrational choices between lotteries, which are explained by their aversion to uncertainty when probabilities cannot be assessed. In such situations, there is a tendency to put higher weights to worst possible outcomes.

Almost all theoretical and empirical research on economics of integrated water management has been dealing with situations characterised by risk, such as stochastic pollution from emission sources or response by aquatic ecosystems to changes in human activities. A common approach for identifying environmental ambition levels and optimal policy instruments has been stochastic programming, where cost effective solutions have been identified for both load and probability constraints, i.e. targets have been formulated for achieving certain load reductions or quality achievements at a minimum probability. A common result from these studies is that costs for achieving certain quality targets increase with higher reliability requirements due to the need of more costly actions for assuring the target. This is illustrated for nitrogen and phosphorous costs presented in Figure 2, where reliability was not of concern. When reliability is important, and the minimum acceptable probability of achieving the different levels of reductions shown in Figure 2 is 0.95, total costs increase as shown for nitrogen reductions in Figure 5 for two minimum probability levels, pr=0.8 and pr.=0.975

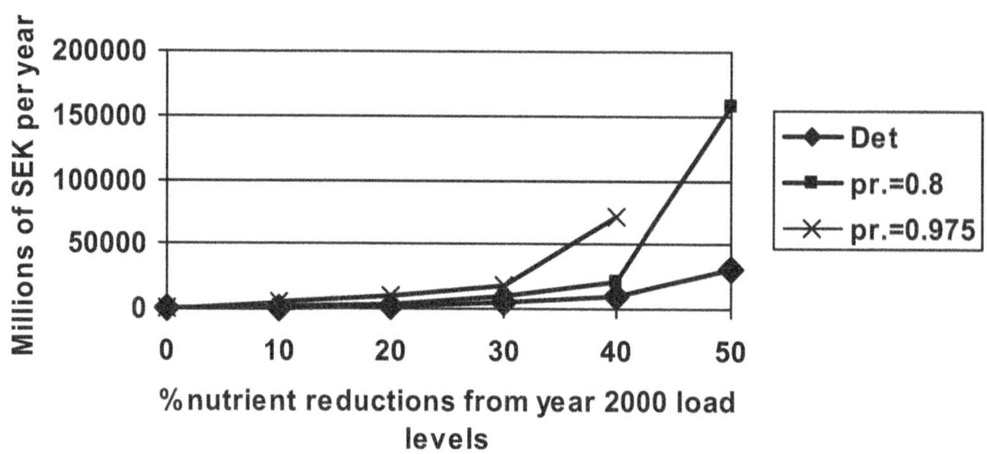

Figure 5: Minimum costs for nitrogen reductions under deterministic and stochastic pollution with reliability constraint of 0.8 and 0.975. (BDBP in appendix)

The curve Det. shows cost for different nitrogen reductions under deterministic conditions, and the two other curves present minimum costs for different nitrogen reductions when the minimum acceptable probability of achieving the load targets is either 0.8 or 0.975. At the higher probability level, reductions exceeding 40 per cent are not possible. The results show how costs increase rapidly for both probabilistic constraints at 30 per reduction requirement, and also that they are considerably higher for larger reliability constraints.

Another result is that measures which affect both mean and variability, such as construction of wetlands, have a cost advantage to measures which affect only the mean, usually improved cleaning at sewage treatment plants. The EU funded ECOWET project resulted in an interesting observation, which points at the role of a portfolio allocation of measures depending on their co-variations (Byström *et al.* 2000). For example, total risk in nutrient load to a coastal ecosystem is reduced when the covariance between upstream leaching and downstream wetland abatement is positive, which means that the abatement capacity is high when leaching is high. This can also be of importance for large-scale transboundary water management if countries' nutrient loads and effects on the aquatic system show covariations. A study of the Baltic Sea showed that countries' risk aversion and covariation in nutrient loads can have significant impacts on management strategies and associated costs and benefits (Gren and Folmer 2003).

Stochastic programming was also applied within the framework of the EU funded project CLIME, which calculated costs under different assumption of climatic changes in the drainage basin of lake Mälaren, located close to the Swedish capital Stockholm (Gren *et al.* 2007b). It constitutes an early application of this method to integrated water management at a relatively large drainage basin scale, with several water quality targets in different lake basins and inclusion of both nitrogen and phosphorous. Climatic change impacts were then assumed to affect both mean and variability in nutrient loads to the lake Mälaren and in the biological response of the lake to nutrient load changes.

Policy design under risky conditions has also been approached from another angle, namely under asymmetric information where involved stakeholders have different information. Segerson (1988) is a pioneering contribution within this field, who suggested a mechanism where polluters in a drainage basin react on their common impact on quality changes in a water recipient. For example, if it is found that water quality is below a targeted value, charges are increased for all polluters in the drainage basin and vice versa. The mechanism has an advantage of creating incentives for people to collect information on the relation between polluting activities and water quality. This research has been followed up and applied to agricultural water pollution in Europe by, among others, Hansen (2002) and Hansen and Romstad (2007). However, the project search on Cordis did not reveal any such EU funded project.

Although relatively many studies have been investigating water management strategies under risky conditions, there are very few considering uncertainty, or ambiguity, aversion. The only study on water management in Europe is Roseta-Palma and Xepadadeas (2004) who apply robust control on pumping of ground water. When managers are ambiguity averse and put relatively high weights to the worst possible outcome, the results justify the precautionary principle usually applied in environmental policies.

Methods developed within economics of both risk and uncertainty management have been applied for calculating strategies and gains from learning by doing, where learning is made by past experiences. However, there is a lack of applications to Europeans aquatic systems, and almost no EU funded project has been using such methods for designing and analysing alternative water policies and institutions. A partial approach is provided by the EU funded project HARMONIRIB, which develops tools for quantifying uncertainty in water management decision making, which is a necessary but not sufficient component for identifying institutional design for water management under risk and uncertainty.

5.6 CONCLUSIONS AND DISCUSSION

In this chapter the review of economics analysis of water management has been classified into four different fields; ambition level, policy design, transboundary water management, and risk and uncertainty. The brief review indicated that there is relatively much research within economics in all these fields, but the EU funded research has to a large extent been focused on projects within the first field, costs and benefits from water management. Within this field, pioneering EU funded research has been made of social net benefits from large scale cleaning programs of the Baltic Sea. The Baltic Drainage Basin Project (BDBP) calculated cost effective solutions for nutrient reductions to the Baltic Sea from all countries in the drainage basin and from several sectors (agriculture, sewage treatment plants, industry, and transports) by means of mathematical programming. Based on this model cost effective design of different policy instruments – charges and permit markets – were made. Furthermore, large-scale contingent valuation studies of improved water quality (reduced eutrophication) were carried out for Sweden and Poland, the results of which were transferred to the other countries in the drainage basin. This was the first study ever applied to such a large sea as the Baltic Sea, and also pioneering in including several sources of nutrient pollution.

Other EU funded projects, such as the ECOWET and EVALUWET have also been forerunners with respect to valuation of large-scale aquatic systems. Although the ultimate purpose of ECOWET project was not to identify water management strategies, the focus on wetlands also generated valuation of wetlands as nutrient sinks in drainage basins. This was among one of the first attempts to estimate values of wetlands as pollutant sinks. This also included the hedging value of wetlands for stochastic pollution of downstream water recipients. The portfolio aspect of water management was also considered in the CLIME project, but the economics focus in this program was of much less significance than in BDBP and ECOWET. The ERMITE cost effectiveness analysis is an early example of application to mines and metal pollutants, and the CLIME cost effectiveness study is relatively unique with respect to its inclusion of several pollutants in a stochastic framework.

However, there are few projects within the other three fields of economic research on water management in particular on transboundary water management and risk and uncertainty. Considering that large scale and uncertainty are characteristics for most integrated and adaptive water management situations, this lack of EU funded projects seems to be in urgent need of focus. The very few EU funded transboundary water policy projects are performed relatively early, within the 3rd framework programme. Considering that economic research on transboundary pollution has developed rapidly since late 1990s, there are now a number of methods relevant for applications on EU transboundary waters, which are useful for, among others, identifying possible mechanisms for sustaining agreements on water management. Similarly, there has been an impressive growth in research on economics of water management under risk situations, which has not been accompanied by EU funded research. On the other hand, economic research on environmental management under uncertainty, when probability distributions for different outcomes cannot be found, is an under-investigated field generally. So far, most of the developed theories on decisions under uncertainty are applied on other fields within economics, such as finance and macroeconomics. Since adaptive management under uncertainty, including learning, is a central theme in integrated and adaptive water management there is a strong need for relevant applications in the near future.

Thus, the brief review of economics of water management in a European perspective can be summarised into four main tentative conclusions. The first is that EU funded research has been pioneering and successful with respect to valuation of water systems, which can be explained by the relatively large number of projects of this kind. Another tentative conclusion is that EU funded research seems to lag behind with respect to the ongoing development in environmental economics research on water policies and water management under risk carried out mainly by research scholars in Europe and USA and there is thus a need for application of these methods in a European context. The third conclusion is that economics of transboundary environmental management in general lacks applications to international water bodies although EU funded research has been pioneering with an early application of this kind to the Baltic Sea. The fourth and final main indication from this review is that management under uncertainty, where probability distributions cannot be assigned to different outcomes, is a serious knowledge gap generally and EU funded research thus shares the challenge with the field environmental economics of building up theoretical and empirical tools for large scale water management under uncertainty.

5.7 REFERENCES

Barrett, S. (2005) The theory of international environmental agreements. In Mäler K-G. and Vincent, J. (eds.). Handbook of environmental economics. Volume 3. North Holland, The Netherlands.

Bateman, I., Day, B., Dupont, D., Georgiou, S., Matias, N., Morimoto, S., and Subramanian, L. (2006) Does phosphate treatment for prevention of eutrophication pass the benefit-cost test. CSERGE Working paper EDM 06-13. University of East Anglia, Norwich, UK.

Bateman, I., Cooper, P., Georgiou, S., Navrud, S., Poe, G., Ready, R., Riera, P., Ryan, M. and Vossler, C. (2005) Economic valuation of policies for managing acidity in remote mountain lakes: Examining validity through scope sensitive testing. Aquatic Sciences 67:274-291

Bayramoglu, B. (2006) Transboundary pollution in the Black Sea: Comparison of institutional arrangements. Environmental and Resource Economics 35:289-325.

Brady M. (2003) Managing agriculture and water quality: four essays on the control of large-scale nitrogen pollution Acta Universitatis agriculturae Sueciae. Agraria vol. 369

Brännlund, R., and Gren, I-M. (1999) Costs of Uniform and Differentiated Charges on a Polluting Input: An Application to Nitrogen Fertilizers in Sweden. Topics in Environmental Economics, Ed. M. Boman. R. Brännlund. and B. Kriström. Edward Elgar, UK.

Byström, O., Andersson, H., and Gren, I-M. (2000) Economic criteria for wetlands as nitrogen sinks. Ecological Economics 35:35-45.

Ellsberg, D. (1961) Risk, Ambiguity, and the Savage Axioms, Quarterly Journal of Economics 75: 643-669.

Elofsson, K. (1999) Costs for reducing agricultural nitrogen load to the Baltic Sea. I Topics in Environmental Economics, Ed. M. Boman. R. Brännlund. and B. Kriström. 33-49.

Elofsson, K. (2003) Cost effective reductions of stochastic agricultural in nitrogen loads to the Baltic Sea. Ecological Economics 47(1): 13-31.

Gren, I-M. (1993) Costs for alternative nitrogen reduction policies in the Mälar region. Ecological Economics 7:159-172.

Gren, I-M., Elofsson, K., and Jannke, P. (1997) Cost effective nutrient reductions to the Baltic Sea. Environmental and Resource Economics, 10(4): 341-362.

Gren I-M. and Folmer.H. (2003) International cooperation of stochastic environmental damage: The case of the Baltic Sea. Ecological Economics, 47(1): 31-43.

Gren, I-M., Svensson, L., Carlsson, M. and Bishop, K. (2007a) Efficient and second best policies for management of heterogeneous landscape components. Land use Policy, in revision,

Gren I-M., Petterson, K. and Blenckert, T. (2007b) Climate change and cost effective water quality management in the Swedish lake Mälaren. Climatic Change, in revision.

Hansen, L.G. (2002) Regulation of Non-Point Emissions: A Variance Based Mechanism. Environmental and Resource Economics 21(4): 303-316.

Hansen, L.G. and Romstad, E. (2007) Non-point source regulation - A self reporting mechanism, Ecological Economics, in press.

Knight, Frank H. (1971) [1921] Risk, Uncertainty, and Profit, with an introduction by George J. Stigler. Phoenix Books. Chicago: University of Chicago Press.

Kolstad, C. and Toman, M. (2005) The economics of climate policy. In Mäler K-G., and Vincent, J. (eds.). Handbook of environmental economics. Volume 3. North Holland, The Netherlands.

Markowska, A., Zylicz, T., and Gren, I-M. (2000) Policy instruments and cost sharing of Baltic sea cleaning programs. In Gren, I-M., Wulff, F., and K. Turner (eds.). Managing a sea: the ecological economics of the Baltic. Earthscan, London, UK, pp. 81-95.

Mäler, K-G. (1990) International environmental problems. Oxford Review of Economic Policy 6: 80-108.

Perman, R., MaY., McGilvray, and Common, M. (2002) Natural resource and environmental economics. Third edition. Pearson Education Limited, New York.

Roseta-Palma, R. and Xepapadeas, A. (2004) Robust control in water management. Journal of Risk and uncertainty 29(1): 21-34.

Scharin, H. (2005a) Comparing two approaches of estimating costs of uniform and spatially differentiated policy instruments. I Scharin, H., Management of eutrophicated coastal zones. Inst. för Ekonomi, SLU, Uppsala, Agraria vol. 503.

Scharin, H. (2005b) Policy instruments in the presence of spatial heterogeneity. I Scharin, H. Management of eutrophicated coastal zones. Inst. för Ekonomi, SLU, Uppsala, Agraria vol. 503.

Segerson, K. (1988). Uncertainty and incentives for non-point pollution control. Journal of Environmental Economics and Management 15:87-98.

Söderqvist, T. (2000) The benefits of a less eutrophicated Baltic Sea. In Gren, I-M., Wulff, F., and K. Turner (eds.). Managing a sea: the ecological economics of the Baltic. Earthscan, London, UK, pp. 66-76.

Söderqvist, T., Mitsch, W.J. and Turner, R.K. (2000) Valuation of wetlands in a landscape and institutional perspective. Special Issue, Ecological Economics 35(1).

Turner, R.K., Paavola, J., Cooper, Ph., Farber, S., Jessamy, V., Georgiou, S. (2003) Valuing nature: lessons learned and future research directions. Ecological Economics, 46: 493-510.

APPENDIX: LIST OF INCLUDED EU PROJECTS

sorted by acronyms when available, otherwise name of principal investigator

AGROIWATECH (Cost-effective technologies for wastewater treatment and waste biodegradation in agro-industries with reclamation of resources), European Commission 5th framework programme, Project reference no. ICA2-CT-2002-10010

AQUAMONEY (Development and Testing of Practical Guidelines for the Assessment of Environmental and Resource Costs and Benefits in the WFD), European Commission 6th framework programme, Project reference no. 22723.

BDBP (Baltic Drainage Basin Project), the European Commission 3th framework programme, Project reference EV5VO183.

CHESS (Climate, hydrochemistry and economics of surface water systems), the EU 4th Framework Programme, Project reference no. ENV4970440.

CLIME (Climate and Lake Impacts in Europe), European Commission 5th framework programme, Contract No. EVK1-CT-2002-00121.

COAST (Coastal monitoring and management), the European Commission 5th framework programme, Project reference no. 43.

ECOWET (Ecological-economic analysis of wetlands: Functions, values and dynamics), the European Commission 4th framework programme, Project reference ENV4960273.

EMERGE (European mountain lake ecosystems: regionalisation, diagnostics & socio-economic evaluation), the European Commission 5th framework programme, Contract No. EVK1-CT-1999-00032.

EVALUWET (European valuation and assessment tools supporting wetland ecosystem legislation), the European Commission 5th framework programme, Contract No EVK1-CT-2000-00070

ERMITE (Environmental Regulation of Mine Water in the European union.), the European Commission 5th framework programme, Contract No EVK1-CT-200-0078.

GOUVERNE (Guidelines for the organisation, use and validation of information systems for evaluating aquifer resources and needs), the European Commission 5th framework programme, Project reference no. EVK1-CT-1999-00043

HARMONIRIB (Harmonised techniques and representative river basin data for assessment and use of uncertainty information in integrated water management), the European Commission 5th framework programme, Project reference no. EVK1-CT-2002-00109

IMPRINT-NET (IMplementing Pricing Reforms IN Transport – NETworking), the European Commission 6th framework programme, Project reference no. Pricin/006293/2004.

Pearce, D. (Benefits transfer and economic valuation of environmental damage in the European Union with special reference to health), the European Commission 4th framework programme, Project reference no. ENV4960234

Post, J. (Standards on nitrate in the European Community: Processes of change in policy instruments and agriculture), the European Commission 3th framework programme, Project reference no. EV5V0155.

Swanson,(a) T. (An Economic Approach to the Problem of Toxic Waste : A Theoretical and Case Study Evaluation of Alternative Regulatory Instruments). Project reference no. EV5V0382.

Swanson,(b) T. (Integrated water management in Cyprus: economic and institutional foundations, the European Commission 4th framework programme, Project reference ENV4980788

SWIFT-WFD (Screening method for Water data Information in support of the implementation of the Water Framework Directive), the European Commission 6th framework programme, Project reference no. 502492.

TRABOREMA (Concepts for integrated transboundary water management and sustainable socio-economic development in the cross border region of Albania, Former Yugoslav Republic of Macedonia (FYROM) and Greece). The EU 6th Framework Programme, Project Reference no. 509177

WADI (The sustainability of European irrigated agriculture under water directive and agenda 2000), the European Commission 5th framework programme, EVK1-CT-2000-00057

WATERWEB (Water Resource Strategies and Drought Alleviation in Western Balkan Agriculture), the European Commission 6th framework programme, Project reference no. 509163

6

Vulnerability and exposure to shocks and stresses in river basins: a review of EU research and some avenues for the future

David Sauri

6.1 INTRODUCTION

The subject of this chapter is vulnerability to water-related hazards, its multiple meanings and the ways the concept is used in various EU-funded research (and also development) projects dealing with hazards. In the risk and policy literature vulnerability is a multidimensional concept, which is often defined in disparate and even contradictory terms. In fact, definitions of vulnerability proliferate so much that a website (www.vulnerabilityNet.org) has been created to compile contributions on this matter. The objective of the chapter is therefore to present different views on vulnerability, especially those of an integrated character, and to contrast these views with predominant uses of the concept in EU-funded research and development projects. In particular, the chapter will use the six attributes constructed by Downing *et al.* (2006), and other recent research to describe how vulnerability is used in a number of selected EU-funded projects This review may be especially relevant for integrated approaches to be developed in the future and reflecting how risks and hazards have progressed from being treated simply as technical questions to being treated as social questions as well.

The chapter is organised as follows. First, I discuss the different meanings of vulnerability particularly those emphasizing a social and political component in an integrated fashion. Second I present and evaluate a number of EU-funded projects dealing with water related hazards as to their vulnerability views (if any). Third I focus more explicitly on seven of these projects (those appearing to be more interested in vulnerability) and assess them according to the attributes developed by Downing *et al.* (2006). Finally, in the conclusions I summarise the main points of the previous sections, especially the presence of integrated views of vulnerability in EU-funded projects on water related hazards, and advance some proposals for future EU-sponsored research on this subject matter.

6.2 THE CHANGING MEANINGS OF VULNERABILITY

As said in the beginning, vulnerability remains elusive to a common and accepted definition. In part, this reflects the changing objectives and interests of the risk and hazards field which in the last decades has undergone an important shift in focus: from a technical and managerial perspective in which the most relevant factor was the natural or technological extreme event and its behaviour, to a more social perspective in which the state (broadly defined) of a given society, community, household, etc. appeared as the key issue. Hence, definitions of vulnerability have changed accordingly. Galderisi *et al.* (2005) trace a history of the concept that begins in the 1970s with a technical and managerial view that identifies vulnerability with (physical) exposure and expected damages. Expected damage and actual damage are, however, different things, and in the 1980s vulnerability began to receive more sophisticated analyses which emphasised first, the multifaceted nature of the concept, and second, the need to discuss very carefully the contexts within which damages occur. In the following decade and afterwards, the social dimension of vulnerability gained presence in academic circles, especially after a seminal book on Natural Disasters published by geographers, sociologists and development experts (Blaikie *et al.* 1994). In this book, we found a definition of vulnerability that encapsulates perfectly the alternative view. Susceptibility to be damaged is widened to incorporate other aspects such as the incapacity to anticipate, to face, to withstand, and to recover from the impact of a calamitous event (Blaikie *et al.* 1994; Galderisi *et al.* 2005). In other words and as Pelling (2001 p.5) observes, vulnerable people are those unable to avoid or unable to absorb potential harm.

Therefore and amidst the large and growing number of definitions of vulnerability found in the literature, two broad views can be discerned. A first view contemplates vulnerability from the standpoint of the biophysical event and defines the concept mostly in terms of expected damages. In this approach, human and social aspects are often subsumed under the black box of "exposure" to the point that exposure becomes frequently interchangeable with vulnerability. Most of the natural sciences and technological approaches to hazards adhere to this view, which is also clearly dominant in EU funded research, as we will see later. Alternatively, the second view does not ignore the physical event but postulates that vulnerability must be defined also from the standpoint of society. The "black box" of exposure is opened to reveal differential patterns of vulnerability according to a wide array of factors, and to show possible adaptive conditions that enhance security and therefore reduce vulnerability even in cases where exposure may be large. As said, both views are the product of different research traditions, the first anchored in the natural hazards and epidemiology studies, and the second, and more recent, from the poverty and development literature (Downing *et al.*, op cit, p. 1). Steps towards the integration of both views are possible and desirable, and several theoretical and methodological initiatives have attempted to develop an integrated assessment of vulnerability that, above all, needs to rework especially the social component of the concept. In the following paragraphs I outline some of the most relevant research questions of possible interest for the integrated view.

First of all, the report entitled "Flood Damage, vulnerability and risk perception- challenges for flood damage research" (Messner and Meyer 2005) included in the FLOODsite project (see www.floodsite.net) states that the change of paradigm away from strict flood control measures to broader flood management actions involves a reconsideration of many concepts related to hazards, risks, and floods in particular. Integrated flood management in particular requires "…a better understanding of the interrelations and social dynamics of flood risk perception, preparedness, vulnerability, flood damage and flood management". This report includes a useful revision of the vulnerability concept that the authors summarise in three strands. First, vulnerability is defined broadly as the potential to be harmed by a hazardous event. Generally speaking, an element at risk of being harmed is more vulnerable, the more it is exposed to a hazard and the more it is susceptible to its forces and impacts. This definition pretends to integrate the following views of the vulnerability concept, which the authors detail in a footnote on page 3:

(1) Exposure to biophysical hazard: This is the most common definition of vulnerability. Essentially it involves an analysis of hazardous conditions, of the human occupancy of hazardous areas, and of the characteristics and impacts of hazardous events. The emphasis of research is strongly biased towards the geophysical event while the human dimension is somewhat relegated to a secondary place.

(2) Social content of hazard: This view emerged after the seemingly failures of the scientific/technocratic approaches to curb losses from hazards, especially in developing countries. In this case, communities link vulnerability to coping responses by, including resistance and resilience to hazards, while the biophysical event is placed on a secondary position.

(3) Combination of exposure and social content: Finally, this is a more synthetic approach that attempts to consider simultaneously the exposure dimension and the social conditions and responses. This is the view preferred by these authors although there are very few examples of its application in Europe.

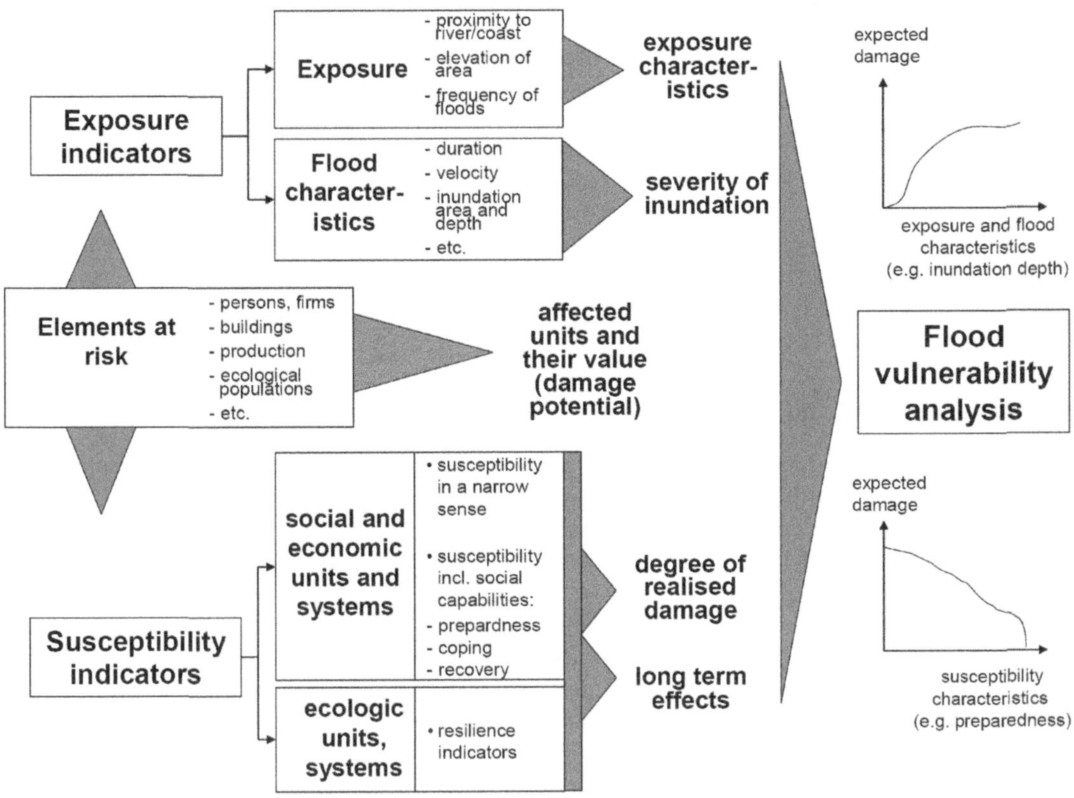

Figure 1: An Integrated View of Vulnerability (Mesner and Meyer 2005, p.4)

As said before, the authors provide three interpretations of vulnerability of which the third one (combining natural and social indicators) is selected. In order to proceed with this interpretation, information is needed about the three key components of vulnerability: Elements at risk, exposure, and susceptibility (see Figure 1). According to the authors, any vulnerability analysis must first take into consideration the elements that may be harmed (that is, humans, properties, ecosystems, etc; see center of fig 1). Damage potential (in both monetary and non-monetary terms) can then be calculated. These calculations provide however, an incomplete assessment of vulnerability since each element potentially at risk may be more or less exposed (in the physical sense) or more or less susceptible (in the social, economic, political, cultural sense) to the natural event.

Exposure indicators are differentiated between those relative to the location (and its physical characteristics) of the element with regard to floods with different return periods, and those relative to the event itself such as duration, depth of floodwaters, velocity of propagation, etc.

More interesting for the purposes of assessing social vulnerability are the susceptibility indicators that are proposed by the authors. These indicators pretend to measure the behaviour of elements at risk when confronted with the actual extreme event. This behaviour may depend first on physical and/or biological conditions. In other words, of attributes of the element at risk that increase or decrease susceptibility; for example age, gender, physical or mental handicaps, etc. or, at another level, type of housing or construction materials. Second, however, behaviour may also be influenced by what the authors call "system characteristics", including the "social context of flood damage occurrence" (p. 5). The latter represents indeed the gist of a more integrated approach to floods and other hazards since it deals with crucial issues and, more importantly in our view, understands hazards and risks (or shocks and stresses) not as isolated events of a certain duration and spatial extent, but as processes that include several stages. Hence, consideration must be given to awareness and preparedness of the population potentially affected by the extreme event; to the capacity of coping with the event, and finally, to the capability to absorb consequences and recuperate once the event has occurred. Indicators for each of these stages are then proposed both at the level of individual households, and at the community level. Many of the indicators in the three stages (for example, availability of emergency services or capacity to purchase flood insurance) depend strongly on socioeconomic conditions such as the financial reserves of the affected, the level of external support by relatives, friends, the state or private donors; the degree of community cohesion, and the political influence to obtain resources for hazard management at the local level, etc.

Regarding susceptibility, the authors also introduce the concept of resilience (p. 6, footnote), which is treated as an antonym of vulnerability. Following the well-known definition of Holling, social resilience is defined as the ability of groups and communities of dealing with external perturbations without suffering substantial negative consequences. Even closer to Holling's original argument, ecological resilience is defined as the ability of a system to absorb external (negative) change. Here the authors refer to some negative impacts of flooding on ecosystems (pollution and excess sedimentation).

The identification and quantification of indicators should be followed by the detection and analysis of the most important relationships between damages and the exposure and susceptibility characteristics of the systems under stress. This step takes the form of functions such as those depicted in the right part of Figure 1. The conceptual apparatus developed by these authors justifies, according to them, a more comprehensive definition of vulnerability "...as the characteristics of a system that describes its potential to be harmed". These characteristics are best expressed as functional relationships of the sort of those depicted in Figure 1; that is, relationships between damages and exposure, and relationships between damages and susceptibility.

6.3 THE SOCIAL DIMENSION

From the social angle, the main question for vulnerability researchers is not so much what are the characteristics of the causal agent but rather, what are the different factors (economic, social, political, cultural, environmental) that influence the endangerment of persons, households, communities, etc. Along this line, other questions asked are, for example, the conditions that influence security (used often as an antonym of vulnerability, see Hewitt, 1997) or the factors behind the ability to recover from the calamity (Blaikie *et al.* 1994, Bohle, 2001). Hence, a formal definition of vulnerability would be "an aggregated measure of human welfare that integrates, environmental, social, economic, and political exposure to a range of harmful perturbations" (Bohle *et al.* 1994, cited in Downing and Ludeke, 2002). Physical exposure is never ruled out of the equation but does not necessarily appear as the main component. In fact, issues such as the individual and societal capacity to prevent or absorb the impacts of the event (in other words individual and communitarian security mechanisms) may be more relevant. In Figure 2, vulnerability is represented as of having three dimensions located at the vortexes of the triangle (exposure, potentiality, and capacity). In turn, these dimensions are governed by conditions of production and exchange and, at a more fundamental level, by class relations and empowerment, by endowments (of, say, a household), and by the specific relationship with the natural world (political ecology).

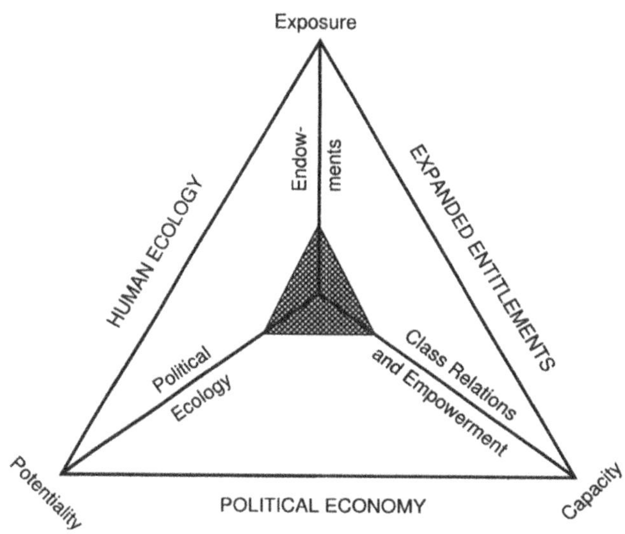

Figure 2: An integrated view of vulnerability (social perspective) (Source: Bohle *et al.* 1994, reproduced in Downing and Ludeke, 2002)

Comparing this alternative view with the more conventional view is illustrative for ascertaining the fundamental distinction to be made between exposure and vulnerability. One person or group may live in a hazardous area and therefore become exposed to a natural event but still remain little vulnerable to the event because he or she has the resources (or the social and/or political influence) to invest in prevention or to purchase insurance. On the other hand, another person or community may live in areas less exposed to extreme events but be more vulnerable because they lack the resources necessary to prevent, cope and recover after rarer extreme events.

One important effort of the more social view of vulnerability concerns the opening of the "black box" and the interest in researching the conditions that make people and communities more vulnerable or insecure. Hewitt (1997) observes two basic vulnerability axes: one social, and one territorial or geographical. In both cases the premise is that not everybody or everywhere in the area affected by an extreme event is equally vulnerable. Vulnerability may vary according to biological factors (age, gender, physical, sensory and mental capacities) and also to socioeconomic factors (income, education, access to political power, etc.) in what some authors call "structural vulnerability". It can also vary spatially, depending on the socio-environmental characteristics of the lands affected by the hazardous event (Downing and Bakker, 2000; Downing *et al.* 2006).

Structural vulnerability emerges from more or less permanent socioeconomic conditions. Steady incomes facilitate resources and choices of individuals and communities facing extreme events but economics is not the only factor in vulnerability. Political power or the possibility of influencing public decision making is also very relevant, for example in what concerns investments for risk reduction or mitigation measures. Hence vulnerability can also be framed as inequality in access to political power and to the resources that emanate from this power (Bankoff *et al.* 2004). Many examples around the world give credit to this assertion. The case of New Orleans flooded by Katrina may be a perfect example of how economic poverty and political inability combined to produce situations of differential vulnerabilities. Moreover, in some cases, vulnerability multiplies along class, race, gender and age divides. Thus a poor and elderly woman belonging to a politically powerless racial group may accumulate several vulnerabilities and become extremely prone to the effects of a natural hazard. An interesting summary of the social dimension of vulnerability that includes an attempt to build an index appears in the work of the Centre for Ecology and Hydrology of Wallingford (see CEH, n.d.) and particularly of Caroline Sullivan and Jeremy Meigh. These authors propose a "Climate Vulnerability Index" regarding water based on the weighted average of six variables: resources (physical availability of water); access (physical appropriation of the resource); capacity (people's ability to manage water); use; environment, and geospatial situation (contextual geographical factors).

6.3.1 The geographical dimension

From a spatial or geographical perspective, vulnerability may also take different forms. Obviously there is not a place free from experiencing the effects of an extreme natural event (in the same way that no individual or group holds absolute protection). That said, however, impacts tend to vary even at small and very small geographical scales, so there may be many and not just one geographies of vulnerability. One of such geographies is of course related to physical exposure, as the more conventional view on vulnerability would argue. Other geographical areas may see their vulnerability enhanced or reduced because of other factors. For instance, areas with deteriorating environments (floods in Haiti and the Dominican Republic); places that become marginal for geopolitical or accessibility reasons (the Kashmir earthquake of 2005) or simply defenceless areas that lack adequate preventive measures or may have their collective mechanisms of solidarity eroded for political reasons (drought episodes in the Horn of Africa). Often and as in the social case, these different factors combine to produce what we could call "super" vulnerable territories.

These approaches are progressively becoming more and more used in a scientific and policy literature. For example, the proposed definitions of the Intergovernmental Panel of Climate Change (IPCC) regarding the terms "Sensitivity", "Adaptive Capacity", and "Vulnerability" indicate the shift towards more social issues. Thus for the IPCC, vulnerability to climate change integrates both the natural (or human-induced) causes, the likely impacts, and the social actions related to this issue: *Vulnerability is a function of the character, magnitude and rate of climate change and variation to which a system is exposed, its sensitivity, and its adaptive capacity" (IPCC, 2001, box in p. 21).*

Mark Pelling (2001 has proposed a synthesis between physical and social vulnerability, which encapsulates many of the arguments developed above. In Figure 3 it can be observed how the explanatory order for vulnerability does not begin with the natural event but with the state of human communities. From this side, vulnerability is said to encompass three dimensions: exposure, resistance, and resilience. The concept of exposure is already well-known and we will not insist in explaining its characteristics. The other two concepts have their roots in population ecology (Holling *et al.* 1998). According to Pelling, "resistance" should be understood as the social capacity to continue with normal life after a perturbation. Resistance depends fundamentally on the conditions of daily life, especially those concerning welfare and public health. Resilience, on its part, refers to the capacity of this same community or society of recovering as quickly as possible from a perturbation. In this case, resilience is said to depend on the degree of social preparedness for extreme events and includes all the specific policies, actions and resources addresses specifically to prevent or mitigate these events.

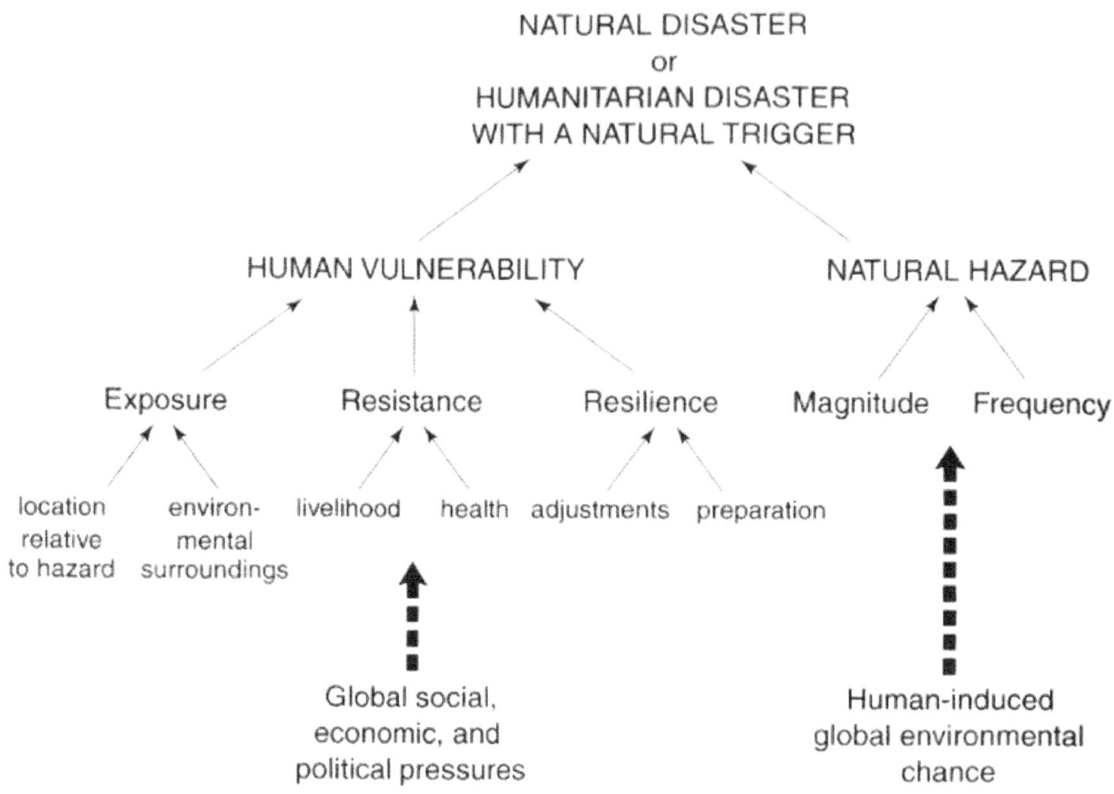

Figure 3: Vulnerability and Hazard: An integration perspective (Pelling 2001 p. 182)

As can be inferred from Figure 3, global economic, social and political pressures and their different effects on different areas of the world may also affect exposure, resistance and resilience. Thus in the developed world, exposure may increase due to the concentration of population and wealth in hazardous areas. Nevertheless, resistance and resilience may attenuate the impacts of the event since, generally speaking, events do not usually imply a substantial alteration of the daily life of the majority of the population. Frequent flash floods in Mediterranean Europe, for example, may cause deaths and disrupt communications but do not pose a threat to most of the population beyond the level of more or less acute annoyances. This picture is, of course, very different in the developing world.

Resilience checks the system's capacity to prevent the effects of extreme events and to mitigate consequences. Also in this case and again speaking generally, despite the fact that the disruption caused and the losses suffered may be considerable, resilience in the developed world tends to be high since developed countries have at their disposal all the means necessary to overcome relatively quickly crisis originated by extreme events. In this sense, some studies even emphasise that, in relative terms, losses caused by natural calamities would have decreased and not increased during the last decades. In other words, the rate of increase in economic losses would lag behind the rate of increase in economic growth (White *et al.* 2001). If material wealth and all the accompanying resources grow faster that losses one could conclude that, at least until now, the growing exposure to hazards characterizing the developed world does not translate *ipso facto* in an increase in vulnerability. To be sure, at scales other than national, situations can be very different. Nevertheless, redistributive mechanisms ensured by states and supra-national organisations such as the European Union could compensate for lesser levels of resilience. In Figure 3 it can also be observed how the natural event is by no means suppressed from the equation, although it loses the determinant character that it usually has in the more conventional approach. Furthermore, it is suggested that extreme natural events may be linked to global environmental change, and especially, climate change originating in human activities.

This multidimensional conceptualisation of vulnerability requires new methodologies as well, especially methodologies that aim at integration of the different components of vulnerability. One of the most important challenges is to develop a vulnerability index able to integrate the many dimensions present in the concept. Such indexes may be quantitative (using for example multivariate statistics) or qualitative (set of subjective values assigned, for instance, by expert knowledge). In this respect, it is also interesting to mention new initiatives in vulnerability mapping that take advantage of the analytical capabilities of Geographic Information Systems (GIS).

Clark *et al.* (1998) sought to map the differential vulnerability to flooding by storm surges in the town of Revere, in the Boston Metropolitan Area. Vulnerability was calculated using a 3x3 matrix, which encompassed on the one hand physical exposure (that is, areas potentially affected by flooding with specified return periods), and on the other hand, intrinsic or structural vulnerability the latter reflected in an statistical index that integrated up to 30 sociodemographic and economic variables obtained from the US Census. The combination between physical exposure and structural vulnerability gave as a result a vulnerability map that identified nine different areas in the city. The map was certainly useful since it identified the areas most likely to require more attention by public bodies (that is, those areas where physical exposure and structural vulnerability where the highest).

Susan Cutter and associates (Cutter *et al.*, 2000) developed maps for multiple hazards occurring at a single place. Taking Georgetown county (South Carolina, USA) as a case study, these authors calculate and map what they call "biophysical vulnerability"(number and frequency of extreme natural events) and "social vulnerability" (from sociodemographic and economic data obtained from the US Census). The two resulting maps are overlaid and the resulting map differentiates up to five different vulnerability categories. As in the Revere case commented above, vulnerability is characterised by a high degree of spatial variability, and areas with high biophysical vulnerability and those with high social vulnerability do not necessarily coincide.

The development of vulnerability maps along the lines of the examples above may be extremely useful for civil defence organisations and emergency planners and direct as well investment priorities towards the more vulnerable communities. These approaches, however, are still plagued with methodological hurdles, for example, the calculation and acceptance of social vulnerability indexes that represent adequately all the variables at hand (provided of course that there exists reliable data on this variables). Another challenge is to abandon vulnerability (which for some authors always bears an implicit bias towards natural events) and focus on the construction of security. In other words, assess first the conditions of daily life and their influence on resistance and resilience factors. Individuals and communities with sufficient levels of security might absorb relatively well extreme phenomena. If this is true, then natural events considered as hazards might turn into events considered as resources; for example, floods that no longer cause destruction and suffering but that improve the quality of soils, replenish groundwater aquifers or contribute to the maintenance of healthy riverine and wetland ecosystems.

As far as integration is concerned, perhaps one of the more comprehensive attempts is that developed by Downing *et al.* (2006, p.3). These authors identify six attributes or characteristics of vulnerability that capture virtually all the main components of an integrated approach:

(1) Vulnerability is the differential exposure to stresses experienced or anticipated by different exposure units.
(2) Vulnerability is a dynamic process, changing on a variety of inter-linked time scales.
(3) Social vulnerability is rooted in the actions and multiple attributes of human actors
(4) Social networks drive and bound vulnerability in the social, economic, political and environmental interactions
(5) Vulnerability is constructed simultaneously on more than one scale
(6) Multiple stresses are inherent in integrating vulnerability of peoples, places, and systems.

The comprehensiveness of these attributes make them suitable to assess the status of vulnerability in a number of EU-funded projects dealing with water and water related hazards as I will attempt to show in the next section.

6.4 VULNERABILITY TO WATER-RELATED HAZARDS IN EU-FUNDED PROJECTS

This review started by preparing a compilation of EU funded projects on water related hazards obtained from the CORDIS and LIFE databases and in the reviews performed by NeWater and FLOODsite projects as well as in the list compiled for the EU-MEDIN network. Between 1994 and 2006, a total of 48 projects with some implicit or explicit reference to vulnerability were identified (see Table 1). The NeWater review identified three projects in which "vulnerability" appeared as a keyword. As said, however, many more included vulnerability either implicitly or explicitly. The projects presented here (see Appendix 1) correspond to the Fourth, Fifth and Sixth Framework Programmes (including some Concerted Actions) and in one case, to an INTER-REG III initiative. Some of them have links to other projects also dealing with vulnerability and exposure relevant for this review. As said, our primary sources of information have been the NeWater report "The Adaptiveness of IWRM; an analysis of European IWRM research" (Timmerman, 2006); the FLOODwater report "Language of Risk. Project Definitions" (Samuels, 2005), and the aforementioned Cordis and Life databases. Most of these projects can be found in the EU-MEDIN network.

A cursory review of Table 1 and Appendix 1 reveals that, to begin with, what we have called the conventional view on vulnerability widely dominates funded research whereas the alternative view remains quite marginal (and therefore underfunded). In the conventional view, the natural event receives absolute priority and the bulk of the research is directed towards the study of its characteristics and behaviour. Socio-environmental systems present in the spatial and temporal frames of the natural event tend to be treated as a "black box". What matters is their presence or not (exposure) in the way of the natural event. Another important feature of this view is the lack of attention given to the pre and post conditions of the calamity since hazards such as floods and droughts tend to be primarily understood as individual and isolated events in space and time.

Table 1. A classification of EU-funded projects on water-related hazards, 1994-2006 (Compiled from lists in the Newater, Floodsite, and Eu-Medin projects as well as from information in the CORDIS and LIFE data bases)

TOPIC	NUMBER	PERCENTAGE
Climate, rainfall-runoff analysis, flood forecasting, and other physical/technical themes:	32	70
Spatial planning and hazard mapping:	3	6
Integration:	2	5
Ecosystems:	3	6
Miscellaneous:	8	16
TOTAL	**48**	**100**

Admittedly, the classification presented in Table 1 may be a bit forced because many projects have more than one specific interest. Hence the relatively high number of the category "miscellaneous". That said, it is worth noting that 70 percent of funded projects have to do with physical/technical issues and for the most part do not mention vulnerability as a possible research focus. Miscellaneous projects (because they attempt to study different aspects of flooding) come next and represent 16 percent of the total. Around 6 percent of the projects refer explicitly to spatial/land use planning and hazard maps, whereas only two (5 percent) have a clear integrative intention. The remaining three (6 percent) have an ecological content.

Hence, we must conclude that EU funded research on this topic is heavily biased towards "hard science" projects that, for the most part, do not include vulnerability (and much less in the social sense) among their primary concerns, although it may be eventually mentioned. This is fairly common (explicitly or implicitly) in projects involving flood forecasting or flood risk assessment such as ACTIF, FRAMEWORK, MICE or RAMFLOOD. One interesting case is the ESPON project (European Spatial Observation Network) which develops and analyzes remote sensing data but that includes a glossary where the definition of vulnerability (as the degree of fragility of human systems) is clearly social. Surprisingly, however, in the executive summary (see www.gtk.fi/projects/espon) vulnerability is interpreted again as physical exposure, and some arguments in agreement of this view are presented, for example that, in Europe, "densely populated areas with a high regional GNP per capita show the highest vulnerability as the total amount of assets per square kilometre poses a higher vulnerability of total damage in case of disaster". Nevertheless, ESPON-HAZARDS has produced the first European wide vulnerability map to natural, technological, and combines hazards, that explicitly include a social dimension (see below).

Most projects do not mention explicitly vulnerability analysis among their objectives. From Appendix 1 it can be observed that 9 out of 48 projects include vulnerability in their presentations, objectives or results. At least two other projects (and possibly more) mention vulnerability in some of their reports (for example EUROTAS). Although vulnerability may not be explicitly mentioned in the main summaries, it may appear somehow in the projects, especially those focusing on spatial planning or hazard mapping. Most of these projects, however, restrict their view to the first of the interpretations stated above (i.e. vulnerability as exposure).

6.5 TOWARDS AN INTEGRATED VIEW ON VULNERABILITY? ASSESSING SEVEN EUROPEAN PROJECTS

Of the 48 projects reviewed, only 5 focus clearly on vulnerability. These are respectively SIRCH, SWURVE, TWINBAS, FLOODsite, and ESPON-HAZARDS. The latter does not mention vulnerability in the title but offers perhaps the most interesting methodological attempt at calculating some sort of vulnerability index that incorporates social data. The INTER-REG III project IRMA-SPONGE does not cover vulnerability per se either but some of the projects included examine issues of vulnerability reduction through the concept of "living with floods". Finally, the ARMONIA project also deals implicitly with vulnerability reduction in that it seeks to evaluate the potential role of spatial planning in this reduction.

Table 2: Attributes of vulnerability in selected EU-funded projects. Projects are ordered according to vulnerability relevance

ATTRIBUTES / PROJECT	Differential exposure	Temporal dynamics	Social Action	Social networks	Geographical scales	Integrating vulnerability
SIRCH	[Very relevant]	[Relevant]	[Very relevant]	[Very relevant]	[Little relevant]	[Little relevant]
FLOODsite	[Very relevant]	[Relevant]	[Very relevant]	[Very relevant]	[Relevant]	[Little relevant]
ESPON-HAZARD	[Very relevant]	[Little relevant]	[Little relevant]	[Relevant]	[Very relevant]	[Little relevant]
SWURVE	[Relevant]	[Little relevant]	[Little relevant]	[Little relevant]	[Little relevant]	[Very relevant]
IRMA-SPONGE	[Little relevant]	[Little relevant]	[Relevant]	[Relevant]	[Little relevant]	[Relevant]
ARMONIA	[Little relevant]	[Little relevant]	[Very relevant]	[Relevant]	[Relevant]	[Little relevant]
TWINBAS	[Little relevant]	[Little relevant]	[Little relevant]	[Little relevant]	[Relevant]	[Very relevant]

Very relevant
Relevant
Little relevant

After this description, and as a way of summarizing the main findings we now assess each of the seven projects selected in terms of their performance regarding the six attributes developed by Downing *et al.*. One important caveat is that some of the projects are still ongoing (FLOODsite, for instance) and that, therefore, the assessment will be made regarding the stated objectives and not the results. The six attributes refer respectively to (a) patterns of differential exposure; (b) the consideration of time changes in the vulnerability patterns; (c) the role of social characteristics and actions; (d) the role of social networks; (e) the consideration of geographical scales, and (f) the consideration of vulnerabilities other than human or social (places, systems). The performance is assessed according to three different possibilities which respectively are "little relevant"; "relevant", and "very relevant". These possibilities reflect the degree to which the project deals or may deal with each of the six attributes of vulnerability

Chronologically, the earliest research project with a comprehensive view on vulnerability was SIRCH ("Societal and Institutional Responses to Climate Change and Climatic Hazards: Managing Changing Flood and Drought Risk") funded by the European Commission's Environment and Climate Programme and carried out between 1997 and 2000. SIRCH published a number of working papers of which the most interesting for the purposes of this report is the piece entitled "A Framework for Institutional Analysis" (Bakker, 1999). Vulnerability is approached from the angle of institutions so that vulnerability to say, flood crises, can be seen as the failure of institutions (sets of rules or norms of behaviour) that govern access to food. Hence, SIRCH was interested in developing vulnerability assessments which departed of the exclusive identification of vulnerability with exposure and raised relevant methodological issues such as the identification of vulnerable groups; the understanding of the vulnerability pattern (or vulnerability profile) as lack of capacity to respond to a perturbations; the compiling of vulnerability indicators, and in sum, the integration of biophysical and socioeconomic determinants of hazardous situations.

The second project of possible interest for our review is SWURVE since it explicitly mentions in its title keywords such as reliability, vulnerability and resilience. However, the focus of the project is on uncertainty of forecasting climate change and the likely extreme events associated with it. Vulnerability is conceptualised as risk of failure of various technological systems and thus appears to depend to a large extent on the capacity of these systems to absorb impacts from extreme events. SWURVE proposes to develop probabilistic estimates of critical outcomes and also to develop a "transferable methodology for the measurement of long term sustainability using statistical measures such as reliability, resilience and vulnerability" (see http://www.ncl.ac.uk/swurve/). These presumably referred to hydrologic and hydraulic systems, and the interpretation appears to be one of safety engineering without taking into account, at least, explicitly, social systems. In all, the vulnerability of technological systems can and must be integrated into the overall assessment of vulnerability. An intriguing question here is whether the authors have in mind that technological systems can also be vulnerable beyond simple exposure to a particular hazard.

Third, the TWINBAS project departs significantly from the rest in that it involves a number of extra-European river basins and its basic concern is to estimate the level of vulnerability for water bodies based on the modelled changes caused by climate and human development. As in the SWURVE project, TWINBAS merits special consideration because it moves vulnerability to the realm of the non-human world. As in SWURVE, concern seems to be mostly with exposure to perturbations, and not with other, more intrinsic dimensions of vulnerability of ecological systems. (see www.twinbas.org).

The fourth and perhaps more comprehensive research project including vulnerability is FLOODsite. This project attempts to understand "specific flood processes and mechanisms and methodologies for flood risk analysis and management at all scales" (see www.floodsite.net, main page). Moreover it also aims at the "...understanding of social vulnerability and flood impacts in order to alleviate risk" (ibid). FLOODsite is one of the rare examples of projects that search for some kind of unified, integrated approach to the study of causes, consequences and potential solutions (and their feasibility) to floods. This task involves first a clarification of common terms, concepts and definitions (with equivalents in several languages) in the hazard and risk field. FLOODsite provides a working definition of vulnerability according to which "…vulnerability…encompasses the characteristics of a system that describe its potential to be harmed". In practice, however, the authors are of the opinion that vulnerability (as well as exposure; both appear to be interchangeable terms) is already captured in the consideration of the consequences of a given hazard episode. Hence, vulnerability is subsumed under the more general term "consequences" and risk is defined by the well-known equation r = p x c by which it equals probability times consequences.

The ESPON-HAZARD project is not a standard research effort but a report that constitutes part of a series of cartographic materials on several issues of interest for the European Union. The resulting report, entitled "The spatial effects and management of natural and technological hazards in Europe" (see www.gtk.fi/projects/espon), offers what is as far as we know the most comprehensive attempt at developing an integrated view on vulnerability to natural and technological hazards developed in the continent. First, natural or technological risk is defined as the product between hazard potential (that is the physical/technological event) and vulnerability, defined as "the degree of fragility". The report combines 15 hazard indicators and weighs them according a Delphi panel formed by experts drawn from other EU projects. Vulnerability is measured according to the following main indicators (table 29, p. 85): national GDP/capita (a proxy for coping capacity), regional GDP/capita; population density; number of tourists; cultural significant sites, and significant natural areas. Regional vulnerability is then defined as the combination between damage potential and coping capacity, the latter measured in terms of GNP/capita. The resulting maps present a coarse but novel view of vulnerability in the continent in which the Eastern part of the continent clearly stands out.

On their part, IRMA-SPONGE and ARMONIA merit consideration in our report because they represent well a family of research, and above all, development projects, that have the objective of reducing vulnerabilities not by modifying the physical event but by modifying human behaviour regarding these events. They both can be seen as the "new wave" in flood management policies that shift the bulk of efforts from to technological modification of natural processes to behavioural modification of our attitudes and actions towards these processes. Concepts such as "living with floods", and the need of harmonizing spatial planning with hazard planning bear a distinctive will to enhance security and thus reduce vulnerability (see www.irnma-sponge.org).

6.6 CONCLUSION

This report has reviewed EU research in vulnerability to water-related extreme events, particularly floods. A total of 48 projects have been selected for analysis. However, only a minority of these appear to have some interest in vulnerability and even fewer target systematically the concept and their different interpretations. Most of the projects that mention or deal with vulnerability follow the dominant view according to which vulnerability equals exposure to the natural event whereas other dimensions are ignored. In fact, we have identified only two projects that offer or promise a more integrated view of the subject, incorporating very especially the human and social dimensions. This dismaying assessment reflects on the other hand the poor state of the social sciences in environmental research as EU funding priorities demonstrate. Nevertheless, vulnerability views are changing and evolving into new approaches that move beyond the simple identification of vulnerability with physical exposure. Integrated water Resources Management and adaptive responses to water challenges such as those pursued by the NeWater project may benefit from a theoretically robust and methodologically improving view of vulnerability that has appeared after several decades of research in the developed world. In this view, what really matters are the conditions of the daily life of people and ecosystems since these conditions will influence to a large extent who is vulnerable to natural and technological hazards and to what degree. Concepts such as resistance and resilience (alongside with exposure) become fundamental to understand vulnerability differentials. At the same time, new cartographies clarify the necessary distinction between exposure and vulnerability and may be of extreme utility for hazard planning and emergency management. We have argued that a consensual definition of vulnerability has not been found in the literature and that it would be probably futile to add another one of our own.

It is in our opinion very important to convey these new sets of theoretical and methodological concerns into the EU decision bodies on research and funding. If this is reasonably achieved, security levels of populations and environments can be enhanced, and natural phenomena that now are mostly considered as hazards can be transformed to resources for the benefit of humans and their environments.

6.7 ACKNOWLEDGEMENTS

I would like to thank all participants in the NeWater seminar held in Lelystad (The Netherlands) in October 2006 for their very helpful comments and criticisms on the first draft of this chapter. Other drafts have benefited enormously from comments by Fons Jaspers and Jos Timmerman. Finally, the work of Tom Downing (and previously of Robert Kates, Gilbert White, Ken Hewitt, and Michael Watts) has inspired much of my thoughts about this subject over the last years.

6.8 REFERENCES

Bankoff G.; Frerks, G. y Hilhorst, D. (eds) (2004) *Mapping Vulnerability. Disasters, Development and People*. London. Eartshcan,

Bakker, K. (ed) (1999): A Framework for Institutional Analysis. Working Paper No 3. SIRCH project. Oxford, University of Oxford, Environmental Change Unit.

Blaikie, P., T. Cannon, I. Davis and B. Wisner (1994) *At Risk. Natural hazards, people's vulnerability, and disasters*. London, Routledge.

Bohle, H. (2001) Vulnerability and Criticality: Perspectives from Social Geography. *IHDP Update* **2**/01, 1-5.

Centre for Ecology and Hydrology (n.d.): "Using the Climate Vulnerability Index to assess vulnerability to climate variations". Wallingford, UK, CEH.

Clark, G.E., Moser, S.C., Ratick, S.J, Dow, K., Meyer, W.B., Emani, S., Jin, W., Kasperson, J.X., Kasperson, R.E. and Schwarz, H.E. (1998) Assessing the vulnerability of coastal communities to extreme storms: the case of Revere, MA, USA. *Mitigation and Adaptation Strategies for Global Change* **3**, 59-82.

Cutter, S.L., Mitchell, J.T. and Scott, M.M. (2000) Revealing the vulnerability of people and places: A case study of Georgetown county, South Carolina. *Annals of the Association of American Geographers* **90**(4), 713-737.

Downing, T.E. and K. Bakker (2000) Drought discourses and vulnerability. In Wilhite, D.E. (ed) *Drought: A Global Assessment*. Vol. II London, Routledge.

Downing, T. and Lüdeke, M. (2002) Social Geographies of Vulnerability and Adaptation. In J.F. Reynolds and D.M. Stafford Smith, *Global Desertification- Do Humans cause Deserts?* Berlin, Dahlem University Press.

Downing, T.E., Aerts, J. Soussan, J.Barthelemy, O., Bharwani, S., Ionescu, C. Hinkel, J. Klein R.J.T., Mata, L., Moss, S., Purkey, D. and Ziervogel, G. (2006) *Integrating Social Vulnerability into water management*. SEI Working Paper and NeWater Working Paper No. 4. Oxford: Stockholm Environment Institute.

Galderisi, A., Ceudech, A. and Pistucci, M. (2005) *Integrated vulnerability assessment: the relevance "to" and "of" regional and urban planning*. Armonia Project.

Hewitt. K. (1997) *Regions at Risk. A Geographical Introduction to Disasters*. London, Longman.

Holling, C.S., F. Berekes y C. Folke (1998) Science, Sustainability and resource management. In F. Berkes y C. Folke (eds): *Linking Social and Ecological Systems.Management practices and social mechanisms for building resilience*. Berlin, Springer-Verlag.

IPCC (2001) *Climate Change 2001: Impacts, Adaptation and Vulnerability*. A Report of Working Group II of the Intergovernmental Panel on Climate Change. Available at http://www.grida.no/climate/ipcc_tar/wg2/index.htm (last accessed 15 December 2006)

Messner, F. and Meyer, V. (2005) *Flood damage, vulnerability and risk perception –challenges for flood damage research*. FLOODsite report. Available at http://www.floodsite.net/html/partner_area/project_docs/DP_1305_Flood_damage_and_vulnerability_Messner_Meyer_2005.pdf.

Pelling, M. (2001) Natural Disasters?. In N. Castree y B. Braun (eds): *Social Nature. Theory, Practice and Politics*. Oxford, Blackwell (pp. 170-188).

Samuels, P. (coord) (2005) *Language of Risk. Project Definitions*. FLOODsite project Available at http://www.floodsite.net/html/partner_area/project_docs/FLOODsite_Language_of_Risk_v4_0_P1.pdf

Timmerman, J.G. (2006) *The Adaptiveness of IWRM; An analysis of European IWRM*. Version 2. NeWater project.

White, G.F., R.W. Kates and I. Burton (2001) Knowing better and losing even more: the use of knowledge in hazards management. *Global Environmental Change Part B: Environmental Hazards* **3**(3-4), 81-92.

APPENDIX 1: SYNTHESIS TABLE OF EU-FUNDED PROJECTS ON WATER-RELATED HAZARDS.

PROJECT TITLE	Web page	Main topic	Vulnerability mentioned.	Vulnerability as exposure	Social vulnerability	Integrated vulnerability
ACTIF	http://www.actif-ec.net/	Reduction of exposure by improvements in flood forecasting;. Also reduce vulnerability by reducing uncertainty	Yes	Yes (implicit)	No	No
AFORISM	http://www.itia.ntua.gr/g/projinfo/2/	(flood forecasting; physical)	No	No	No	No
ANFAS Support	http://www.ercim.org/ANFAS/	Decision system data base to simulate flood events	No	No	No	No
ARMONIA	: http://www.armoniaproject.net	Hazard Maps, spatial planning	Yes	Yes	No	No
CARPE DIEM	: http://carpediem.ub.es	Rainfall forecasting, radar applications	No	No	No	No
CLIFF	http://cordis.europa.eu/data/PROJ_FP5/ACTIONeqDndSESSIONeq112422005919ndDOCeq425ndTBLeqEN_PROJ.htm	Disaster management with Earth Observation data	No	No	No	No
CLOUDMAP2	http://www-research.ge.ucl.ac.uk/cloudmap2/	Cloud properties, climate modelling	No	No	No	No
ECOFLOOD	www.ecoflood.org	Natural flood defences in the Netherlands	No	No	No	No
EFFS	http://effs.wldelft.nl	European flood forecasting system (physical variables)	No	No	No	No
ELDAS	http://www.knmi.nl/samenw/eldas	Soil moisture content for flood and drought forecasting	No	No	No	No
ELLA	http://www.ella-interreg.org/	Flood and spatial planning in the Elbe basin (iin	No	Implicit	No	No
ENSEMBLES	http://www.ensembles-eu.org	Climate modelling (physical)	No	No	No	No
EOLES	http://www.spacebel.be/eoles/	Monitoring of hazardous events from satellites	No	No	No	No
ESPON_HAZARDS	www.gtk.fi/projects/espon	Hazard maps and indicators. Spatial planning with a view on climate change	Yes	Yes	Yes (in glossary of terms)	Some (population and GDP)
EURAINSAT	http://www.isac.cnr.it/~eurainsat/	Rainfall estimation from satellites (physical)	No	No	No	No
EUROTAS	http://www.hrwallingford.co.uk/projects/EUROTAS/	use and hydrological to help to mitigate floods	Not in the general projects but appears in some reports	No (but implicitly yes in some reports		
EU-MEDIN	www.eu-medin.org	Sharing of disaster-related information data	No	No (implicitly maybe: interest in socio-economic impacts of floods	No	No

APPENDIX 1 (continued): SYNTHESIS TABLE OF EU-FUNDED PROJECTS ON WATER-RELATED HAZARDS.

PROJECT TITLE	**Web page**	**Main topic**	**Vulnerability mentioned.**	**Vulnerability as exposure**	**Social vulnerability**	**Integrated vulnerability**
FAME PROJECT	http://www.sadl.kuleuven.ac.be/fame/	Flood modelling and mapping	no	No	No	No
FLOBAR	http://www.geog.cam.ac.uk/research/projects/flobar2/aims/	Floodplain biogeography; role of vegetation in flood retention	No	No	No	No
FLOODAWARE	http://www.lyon.cemagref.fr/projets/floodaware/	Physical models of flooding	Yes	Yes	No	No
FLOOD RELIEF	http://projects.dhi.dk/floodrelief/	Decision support system; integration physical variables for flood forecasting	No	No	No	No
FLOODMAN	http://projects.itek.norut.no/floodman/	Flood forecasting with satellite data, etc. (physical)	No	No	No	No
FLOODSite	www.floodsite.net	Largest project on floods; integrated view	Yes	Yes	Yes	Yes
FLOWS	www.flows.nu	Community adaptation to flooding, public agencies	No	Yes (implicit)	No	No
FRAMEWORK	http://www.diiar.polimi.it/framework/index.htm	Physically-based methods for flood risk assessment,	No	No	No	No
HYDROMET	http://www.cen.bris.ac.uk/civil/research/wemrc/hydromet	Storm flood warning through radar	No	No	No	No
IRMA-SPONGE	www.irnma-sponge.org	Wide array of flood management issues in rivers Meuse and Rhine	No, but maybe some reports	No, but maybe some reports	No	No
IMPACT	www.impact-project.net	Physical failure of dams and flood defence works	No	Implicitly	No	No
MANTISSA	http://www.arp-manchester.org.uk/MANTISSA.htm	Microwave links for rainfall measurements	No	No	No	No
MICE	http://www.cru.uea.ac.uk/cru/projects/mice/	Modelling impacts of climate extremes (physical) but also on socioeconomic sectors such as tourism	No	Implicitly (tourism)	No	No
MITCH	http://www.hrwallingford.co.uk/Mitch/	Flood forecasting and warning, public participation	No	Implicit	No	No
MUSIC	http://www.geomin.unibo.it/orgv/hydro/music/index2.htm	Improve reliability of rainfall estimation by physical means	No	No	No	No
NATURAL HAZARDS	http://natural-hazards.jrc.it/	Provision of risk indicators (before and after crisis) and of hazards maps	No	Implicit	No	No
NEDIES	http://nedies.jrc.it/index.asp?ID=67	Data base on natural and technological hazards	No	Implicit	No	No
ORCHESTRA	http://www.eu-orchestra.org/	Software for multi.-risk management	No	No	No	No
OSIRIS	http://www.ist-osiris.org/	Dissemination of information about floods for stakeholders	No	Implicit	No	No

APPENDIX 1 (continued): SYNTHESIS TABLE OF EU-FUNDED PROJECTS ON WATER-RELATED HAZARDS.

PROJECT TITLE	**Web page**	**Main topic**	**Vulnerability mentioned.**	**Vulnerability as exposure**	**Social vulnerability**	**Integrated vulnerability**
RIMABOD	http://www.hrwallingford.co.uk/projects/RIBAMOD/	Catchment modelling, flood forecasting (physical/technical)	No	No	No	No
RINAMED	http://www.rinamed.net/	Natural hazards in the Western Mediterranean	Yes, some definitions, v as exposure	Yes	No	No
SIRCH	http://www.geo.ucl.ac.be/LUCC/research/endorsed/01-sirch/SIRCH.HTML	Institutions for managing floods and droughts	Yes	Yes	Yes	Yes
SPHERE	http://www.ccma.csic.es/dpts/suelos/hidro/sphere/overview.html	Paleofloods (physical analysis)	No	No	No	No
SWURVE	http://www.ncl.ac.uk/swurve/	Statistical treatment of reliability, resilience and vulnerability	Yes	Yes	No	No
THARMIT.	http://www.unitn.it/events/tharmit/index.htm	Physical characteristics of torrents in the Alps	No	No	No	No
WRINCLE	http://www.ncl.ac.uk/wrincle/	Changes in precipitation after climate change and translation into water resources	No	No	No	No
TWINBAS	www.twinbas.org	Vulnerability for water bodies based on the modelled changes caused by climate and human impacts	Yes	Yes	No	No

7

Understanding consequences of climate change for water resources and water-related sectors in Europe

Fred F. Hattermann, Valentina Krysanova, Joachim Post, Thomas Dworak, Markus Wrobel, Susanne Kadner and Anna Leipprand

7.1 INTRODUCTION

The global mean surface temperature has been persistently increasing over the past decades, and the scientific community came to an agreement shared by most scientists that humankind is witnessing a change in the global climate (IPCC 2007a). The functioning of the earth system is very complex, and there are manifold processes influencing climate, and many systems and processes are influenced by the climate. Therefore, multiple impacts along different causal chains are to be expected worldwide as a result of climate warming.

In Europe, impacts on water resources are among the main concerns. Many recent investigations highlighted the challenges that result from shifts in precipitation patterns and snow regime, changes in seasonal water availability and water quality, rise of sea level, and increase in the frequency and/or intensity of floods and droughts, all coupled with the rise in mean surface temperature. A variety of EU funded projects investigated the possible changes in climate at the global and European scale, the possible impacts on water resources and measures and options for adaptation. Tables 1 and 2 give an overview of European projects of the 5th and 6th Framework Programme, which are related to climate change impact research on water resources (Table 1) and adaptation to climate change (Table 2). The main findings are summarised in different reports assigned by the European Commission, the European Environmental Agency and the United Nations (EEA 2004; 2005a; 2005b; Eisenreich 2005; IPCC 2001; 2007b). This chapter gives a review on the state of the art in climate change research in Europe and indicates based on the investigation possible gaps and issues for further improvement (see Section 6 and 7). In addition to Tables 1 and 2, where the most important projects and their research fields are listed, specific results of these projects are discussed when relevant.

One major result of the reports is that there is no doubt among most scientists that the observed climate warming is to a large extent attributable to anthropogenic activities, mainly to the human-driven increase in emissions of greenhouse gases and change in land use patterns (IPCC 2007a). For that reason the evolution of global emissions of greenhouse gases in future will significantly influence the magnitude of potential climate change and impacts. Therefore the reduction of greenhouse gas emissions should be the primary political goal, especially in industrial countries, in order to mitigate climate change and to prevent most disastrous impacts. The declared aim of the European Union is to limit the temperature rise to 2 degrees above the pre-industrial level (European Commission 2005). This aim has to be translated into concrete policy measures and emission reduction targets.

 The adaptiveness of IWRM; Analysing European IWRM research Edited by Jos G. Timmerman, Claudia Pahl-Wostl and Jörn Möltgen. ISBN: 9781843391722. Published by IWA Publishing, London, UK.

Alongside with the mitigation measures, serious efforts and strategies are also needed to cope with the changes in climate and associated impacts on water resources that are already happening or that are to be expected even in case emission reduction targets will be achieved (Schröter *et al.* 2005). Adaptation to climate change effects has already become an important political issue for developing countries, where some of the effects will be most pronounced. Nowadays, it is also gaining relevance and attention on the European political agenda. The European Commission prepared a "Green Paper on Climate Change and Adaptation", and a working group has been set up early in 2007 that will explore potential adaptation strategies for all relevant sectors in the context of the European water policy. The urgent and crucial scientific challenges currently are to investigate: how pronounced climate change will be in different European regions, which climatic parameters are most sensitive, what impacts on water resources availability and extreme events are to be expected, which mitigation measures are possible, and how society in general and economic sectors in particular can adapt to expected changes.

The following sections describe and discuss the observed trends of climate change over Europe in terms of temperature, precipitation and extremes like floods and droughts, and discuss possible projections for the future (Section 2). Pressures upon water resources induced by other driving forces like land use change and population growth are the topic of Section 3. Section 4 summarises the possible impacts and consequences for water resources and water-related sectors in Europe. Coping strategies now and needed in future are outlined in Section 5, and Section 6 discusses the role of integrated and adaptive water management in coping with consequences of climate change. Finally, Section 7 draws conclusions regarding the state of the art in climate impact research in Europe.

Table 1: European projects of the 5th and 6th Framework Programmes related to climate and land use change scenarios and impact assessment on water resources and ecosystems

PROJECT ABBRE-VIATION	PROJECT NAME	CLIMATE AND WATER RELATED ISSUES	RELEVANCE TO SPECIFIC RESEARCH AREAS*
ACCELE-RATES	Assessing climate change effects on land use and ecosystems: from regional analysis to the European scale	Investigation of the impacts of climate change on land use and biological resources in managed ecosystems in Europe.	2, 3
ALARM	Assessing LArge-scale environmental Risks with tested Methods	Evaluation of combined risks from changes in biodiversity and ecosystem functioning, especially related to climate and land use changes.	2, 3
ATEAM	Advanced terrestrial ecosystem analysis and modelling	Assessment of the vulnerability of major European ecosystem types and services in response to global environmental change.	3
ATLANTIS	Atlantic sea level rise: adaptation to imaginable worst case climate change	Investigation of impacts of sea level rise on society in low-lying representative European regions and evaluation of potential adaptation options.	3, 4
CECILIA	Central and Eastern Europe Climate Change Impact and Vulnerability Assessment	Climate change impacts and vulnerability assessment in targeted areas of Central and Eastern Europe.	1, 3
CLIME	Climate and Lake Impacts in Europe	Assessment of direct and indirect impacts of climate change on European lake systems using indicators of WFD.	3
CLIMED	Effects of climate change variability on water availability and water management practices in western Mediterranean.	Use of statistical and physical models to assess future freshwater availability and develop policy recommendations in response to climate change.	3, 4

Table 1(continued): European projects of the 5th and 6th Framework Programmes related to climate and land use change scenarios and impact assessment on water resources and ecosystems

PROJECT ABBRE-VIATION	PROJECT NAME	CLIMATE AND WATER RELATED ISSUES	RELEVANCE TO SPECIFIC RESEARCH AREAS*
DINAS-COAST	Dynamic and Interactive Assessment of National, Regional and Global Vulnerability	Aggregation of scientific information on coastal impacts of climate change and development of inter-active tool for stakeholders and policy makers to assess coastal vulnerability.	3
EEA OUTLOOK	Outlook on Water Use in Europe	Study prepared on behalf of the European Environment Agency (EEA) as a contribution to the State of the Environment Outlook Report 2005 (SoEOR2005) to produce quantitative estimates of water use in Europe up to 2030.	3
ENSEMBLES	ENSEMBLE-based Predictions of Climate Changes and their Impacts	Development of a common European ensemble forecast system to reduce uncertainties in climate change projections.	1
EUROCLIM	European Climate Change Monitoring and Prediction System	Use of European and Arctic cryosphere as indicators for climate change and establishment of operational monitoring system.	1
EURO-LIMPACS	Integrated Project to Evaluate the Impacts of Global Change on European Freshwater Ecosystems	Investigation of combined impacts of climate and land use changes on freshwater ecosystems and development of toolkit for integrated catchment analysis and modelling.	2, 3
MICE	Modelling the impact of climate extremes	Use of results of global and regional climate models to investigate future changes in the frequency and intensity of extreme events in Europe.	3
MIND	Mediterranean terrestrial Ecosystems and Increasing Drought: Vulnerability Assessment	Study of potential impacts of increasing drought conditions on Mediterranean terrestrial ecosystems and assessment of their vulnerability to future changes in rainfall patterns.	3
PRUDENCE	Prediction of Regional Scenarios and Uncertainties for Defining European Climate Change - Risks and Effects	Use of RCMs, high-resolution GCMs and climate impacts models to project changes at the regional scale and to specify uncertainty in the projections.	1
STARDEX	Statistical and regional dynamical downscaling of extremes for European regions	Development of a systematic framework for identification and evaluation of suitable scenario construction methods to project extreme events.	1,3

*** 1 – climate models and scenarios, 2 – land use change, 3 – impact on water resources and ecosystems, 4 – coping strategies**

Table 2: European projects of the 5th and 6th Framework Programmes related to coping strategies and adaptation to climate change

PROJECT ABBRE-VIATION	PROJECT NAME	PROJECT DESCRIPTION	RELEVANCE TO SPECIFIC RESEARCH AREAS*
ADAM	Adaptation and Mitigation Strategies: Supporting European climate policy	Assessment of mitigation and adaptation potential of existing and evolving EU and world policies and development of alternative strategic policy options.	4
AQUADAPT	Strategic tools to support adaptive, integrated water resource management under changing utilisation conditions at catchment scale: a co-evolutionary approach.	Improvement of knowledge and available tools for strategic water resource planning in semi-arid regions, especially with regard to environmental, land use and socio-economic developments.	2, 4
AWARE	Tool for monitoring and forecasting Available WAter REsource in mountain environment	Development of monitoring and forecasting tools for water resource management in snow-dominated, mountainous environments.	4
GABAR-DINE	Groundwater Artificial recharge Based on Alternative sources of wateR: aDvanced INtegrated technologies and managEment	Identification of alternative water sources and evaluation of their environmental and economic suitability for exploitation.	4
HARMONI-CA	Harmonised modelling tools for integrated basin management	Creation of a discussion forum for modelling community, policy makers and users to evaluate application and development of WFD relevant modelling tools.	4
MEDIS	Towards sustainable water use on Mediterranean islands: addressing conflicting demands and varying hydrological, social and economic conditions	Advancement of more sustainable and equitable use of water in the Mediterranean islands through monitoring of consumption and recharge patterns and involvement of stakeholders to identify better management schemes.	4
MITCH	Mitigation of climate induced natural hazards	Concerted action of researchers and stakeholders to develop possible mitigation strategies and to prepare for climate induced natural hazards.	4
NEWATER	New Approaches to Adaptive Water Management under Uncertainty	Support transition from current river basin water management regime to more adaptive approach of an IWRM concept.	3, 4
RIVERTWIN	A Regional Model for Integrated Water Management in Twinned River Basins	Development of an integrated regional model to be used by water resources management in twinned river basins with contrasting ecological, social and economic conditions.	3, 4
SWURVE	Sustainable water uncertainty, risk and vulnerability in Europe (SWURVE)	Evaluation of probability of projected potential climate impacts on hydrological systems and development of methodology to determine sustainable performance options.	3, 4
TWINBAS	Twinning European and third countries river basins for development of integrated water resources management methods	Assessment of the vulnerability of five twinned river basins to climate change and anthropogenic development and preparation of integrated river basin management plans.	3, 4

*** 2 – land use change, 3 – impact on water resources and ecosystems, 4 – coping strategies**

7.2 CLIMATE CHANGE IN EUROPE: EXISTING TRENDS AND PROJECTIONS

7.2.1 Trends in climate

Key reports summarising the outcome of the European projects listed in Table 1 are IPCC (2007a); EEA (2004) and Eisenreich *et al.* (2005). The result of the studies is that the climate of the earth was never stable, it has always been characterised by certain variability and changes. However, the extent of climate change observed over the last decades and the rate of these alterations exceeds all changes in the last thousands of years (IPCC 2007a; EEA 2004).

Temperature. The increase in average air temperature, observed particularly over the last decades, is one of the most evident signals of global climate change. It is established that the current rate of global warming is about 0.17 ± 0.05 °C per decade (IPCC 2001). The change in temperature is generally more pronounced at higher latitudes. Accordingly, the air temperature over the European continent has increased more than the global average: by 0.8 °C to 0.95 °C since 1900 compared to 0.6 °C as a global average (EEA 2004; IPCC 2001). Figure 1 shows the increase in observed temperature in Europe over the last century. An important phenomenon is that the increase in winter temperature in Europe was higher than that in summer. There are also apparent regional differences: the observed warming has been greatest in northern Scandinavia, northwest Russia, and western Mediterranean (up to 3 °C increase). Other parts of Europe, especially central Europe and the eastern Mediterranean coast, show lower increases in temperature (Figure 1).

Precipitation. A general pattern is that the observed higher temperatures stimulate the global hydrological cycle (more evapotranspiration leads to more water vapour in the atmosphere and to more precipitation). On the other hand, precipitation has a higher regional variability than temperature, because it depends more on regional circulation patterns and local orography. Therefore, trends in precipitation are more heterogeneous (see Figure 1). The observed precipitation trends for the period from 1900 to 2000 show a contrasting picture between increase in northern Europe by 10–40 %, and decrease in some regions of southern Europe with up to 20 % less precipitation, especially in the winter season (IPCC 2007b; Klein Tank *et al.* 2002). Also, according to observations west-wind circulation patterns are becoming more frequent, and precipitation in luff (west part) of mountain ranges increases, while it decreases in lee (east part) of the mountain ranges.

Across the European continent seasonal precipitation patterns show even more distinct trends than the average annual values. In the winter season many parts of Northwest Europe became wetter, and regions in southern and eastern Europe drier. The changes in winter precipitation can partly be explained by stronger western winds over northern Europe, bringing more clouds (EEA 2004). The observed reduction in summer precipitation in the UK, eastern Europe and southern Scandinavia are explained by increases in the duration of high pressure areas in central Europe.

Temperature and precipitation extremes. In consequence of global warming, the number of cold and frost days has decreased in most parts of Europe over the past 100 years, whereas the number of days with temperatures above 25 °C (summer days) has increased (EEA 2004; Jones *et al.* 1999; Klein Tank *et al.* 2002). Important is that the tendency towards milder winters in Northern Europe over the last decades is partly explained by stronger westerly circulation in winters, which is consistent with a positive phase of the North Atlantic Oscillation (NAO) (Hurrell and Van Loon 1997). The positive phase of the NAO index occurs when the difference between the subpolar low and the subtropical high is greater than usual, and it is associated with winds across mid-latitudes pushing storms across the Atlantic on a more northerly track.

In many regions, the trend in precipitation extremes is more pronounced than the trend in average precipitation. Many regions in southern Europe experienced a significant decrease of very wet days (with precipitation > 20 mm) over the last decades, whereas the number of wet days (with precipitation > 1 mm) and the number of very wet days increased in central and northern Europe (EEA 2004; Klein Tank *et al.* 2002). Concerning changes in seasonal precipitation patterns, frequency of heavy rainfall events in winter increased in central Europe, the UK and Scandinavia, and frequency of summer heavy rainfall events increased across northern Scandinavia, south-western Europe and north-western Russia during the last 40 years (according to analysis done in the framework of the STARDEX project, Goodess 2005).

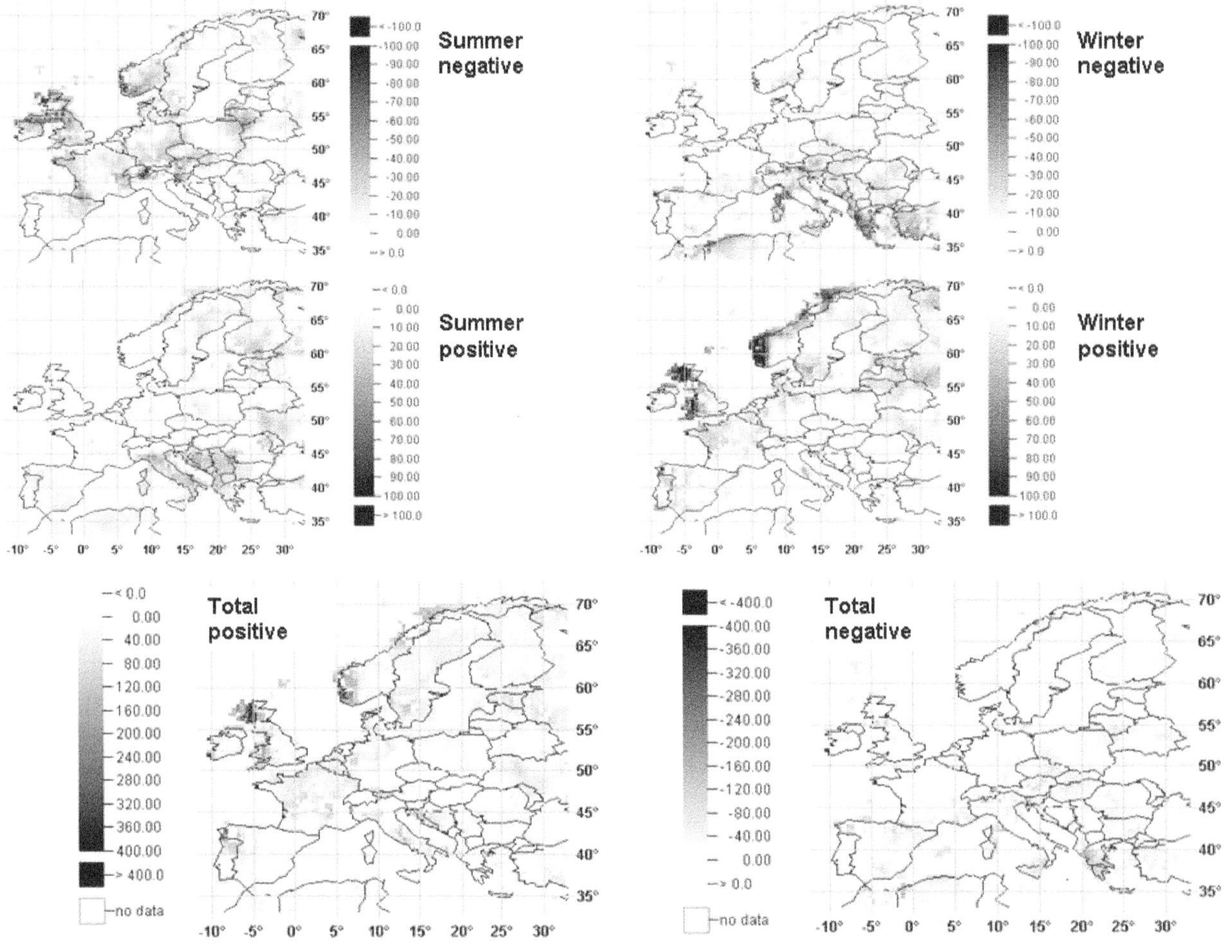

Figure 1: The observed trend in precipitation over the last century (difference mean annual values 1975/2004 and 1931/1960) for the winter (DJF) and summer (JJA) season (seasonal sums in mm), and annual averages (annual sums in mm). (This reference period was chosen instead of 1901/1930, because many time series do not cover the first decades of the 20th century. Source: PIK Global Climate Dataset, 2007, compiled from data of the national weather services).

7.2.2 Climate scenarios and uncertainty

According to the terminology of the IPCC (2001), scenarios are images of the future, or alternative futures. They are neither predictions nor forecasts. Rather, each scenario is one alternative image of how the future might be. When discussing future climate change it is important to understand that there cannot be any prediction of the future. Scenarios are rather *projections* into the future under the assumption that society behaves in a certain way. As a result, climate change varies from scenario to scenario, and is associated with *the scenario uncertainty*.

The second important source of uncertainty is a not fully adequate understanding and/or reproduction of atmospheric circulation processes in General Circulation Models (GCM) (*model uncertainty and parameter uncertainty*). Important is also that the spatial scale of GCMs is currently between 1.0 and 2.5 degrees. Though different GCMs use the same or similar basic physical equations, they produce results, which may vary for the same scenario. Several European projects tackled the problem to improve the quality of climate scenario results at the regional scale in Europe, among them the projects STARDEX, ESSEMBLES and most recently the project PRUDENCE (see table 1 and Figures 2 and 3).

For regional scale impact assessment usually results of Regional Climate Models (RCMs) are used as they provide finer spatial resolution (to avoid uncertainties related to scaling). Two main methods are currently in use: dynamic downscaling methods, basically having the same mathematical background as GCMs, and statistical methods, where the model "learns" from observations under certain boundary conditions in the past to make assessments for the future. RCMs use the results of GCMs as boundary conditions, but they better reflect local topographical and coastline features, which influence climate and local physical processes (especially relevant for precipitation and for the summer period, e.g. convective processes in the atmosphere). Hence, when available and validated for plausibility, RCMs scenarios should be applied for regional and river basin scale impact assessments. The finest resolution currently possible in dynamic RCMs is 5-10 km, whereas statistical methods normally produce scenarios for the net of climate stations where the statistics (correlations) of the climate pattern where derived. The results of the PRUDENCE project shown in Figures 2 and 3 indicate that there is still high uncertainty about climate change in different European regions, especially in central and eastern Europe. However, the trends as such (positive or negative) mostly show the same direction and seasonal pattern.

7.2.3 Expected climate change in Europe according to scenarios

Temperature. In Europe the average temperature is projected to increase by 1.0–5.5°C by 2100 compared to 1990, if no policy measures will be taken (IPCC 2007b). The European target of limiting global temperature increase to no more than 2.0 °C above the pre-industrial level is likely to be exceeded around 2050 (EEA 2004). The range of uncertainty results from different potential pathways of technological, economic and demographic development leading to different emissions of greenhouse gases and related response of the climate system.

The possible changes in temperature under scenario conditions across Europe and the associated uncertainties have been investigated in the PRUDENCE project (Jacop *et al.* 2007). Figure 2 illustrates the projected change in temperature until 2100 as a result of a set of different RCMs. Important is that the relative spatial pattern projected by each climate model remains the same over one emission scenario (A2), and only the size of the anomaly varies between the models. A general pattern is that except Scandinavia, the increase of temperature is the highest during summer and lower during winter. Scandinavia shows a general increase in temperature by about 3-5 °C. The highest uncertainty exists for eastern Europe, especially during summer.

Within Europe, the warming is estimated to be less pronounced along the Atlantic coastline (British Islands).

Precipitation. Projections show a 1–2 % increase per decade in annual precipitation in northern Europe and a decrease up to 1 % per decade in southern Europe (EEA 2004). Most of Europe is likely to become wetter in the winter season (1–4 % per decade) with some exceptions (the Balkan countries and Turkey). Despite the precipitation increases, the amount of snow and area covered by snow are expected to decline due to warming. In summer, northern Europe might become up to 2 % wetter per decade, whereas southern Europe may become up to 5 % drier per decade (Parry 2000).

Figure 3 illustrates the projected seasonal change in precipitation under scenario conditions until 2100 as a result of the same RCMs as in Figure 2 (Jacop *et al.* 2007). As for temperature, the overall pattern of change is similar for different RCMs for the same emission scenario (A2), although the spatial heterogeneity is larger than for temperature. A general picture is that the summer precipitation all over Europe, except Scandinavia, decreases, while in most cases winter precipitation increases. However, the winter increase in precipitation is small in the Mediterranean and cannot compensate the decrease in summer. The highest uncertainties in simulated precipitation trends exist again for East Europe. It has to be mentioned that the FP6 project CECILIA focuses on high resolution climate modelling and impacts assessment in target areas located in central an eastern Europe.

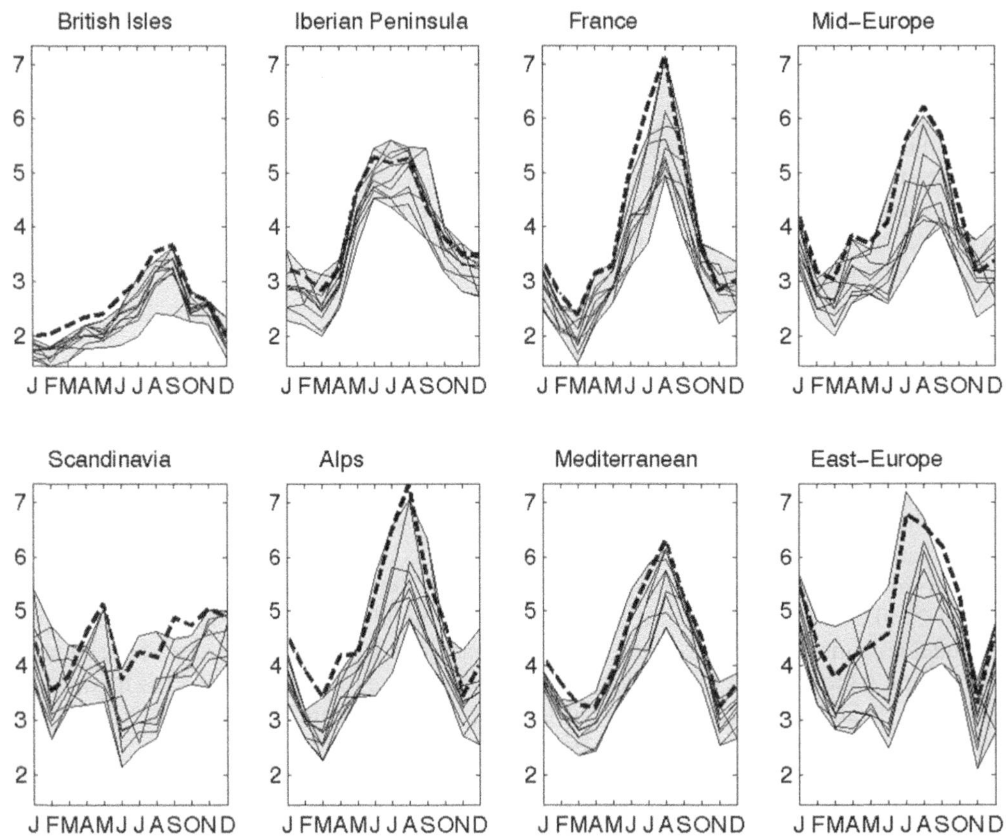

Figure 2: Monthly changes in temperature for different regions in Europe simulated by an ensemble of different RCMs in the project PRUDENCE (Jacop *et al.* 2007) Shown is the signal in 2m-Temperature, 2071-2100 minus 1961-1990 for the A2 SRES scenario. The thick black line indicates the result of the driving GCM for the region.

Temperature and precipitation extremes. According to the report of the European Environmental Agency (EEA 2004), cold winters, which occurred on average once every 10 years in the period from 1961 to 1990, are likely to become rare in Europe and will almost fully disappear by 2080. Besides, by 2080 in many European regions almost every summer is projected to be hotter than the 10 % hottest summers in the current climate. Under high emission scenarios by the end of the 21st century every second summer in Europe will be as hot or even hotter than the exceptionally hot summer of 2003 (Luterbacher *et al.* 2004).

Notable changes are also projected for precipitation extremes in Europe, though the uncertainties in the prediction of extreme precipitation are still high. For example, Palmer and Raisanen (2002) project that the probability of extremely high precipitation over Europe in winter would increase by about two to five times over the next 50 to 100 years, leading to increased risk of severe floods. Other studies argue that it is likely that the frequency of both intense precipitation events and summer drought risk in Europe will increase (Klein Tank *et al.* 2002; Parry 2000). However, the attribution of these extremes to climate change is still uncertain due to a lack of accurate data and a full scientific understanding of the climate system functioning. Some first assessments have appeared showing that extremes are changing more than the average climate (e.g. Schär *et al.* 2004).

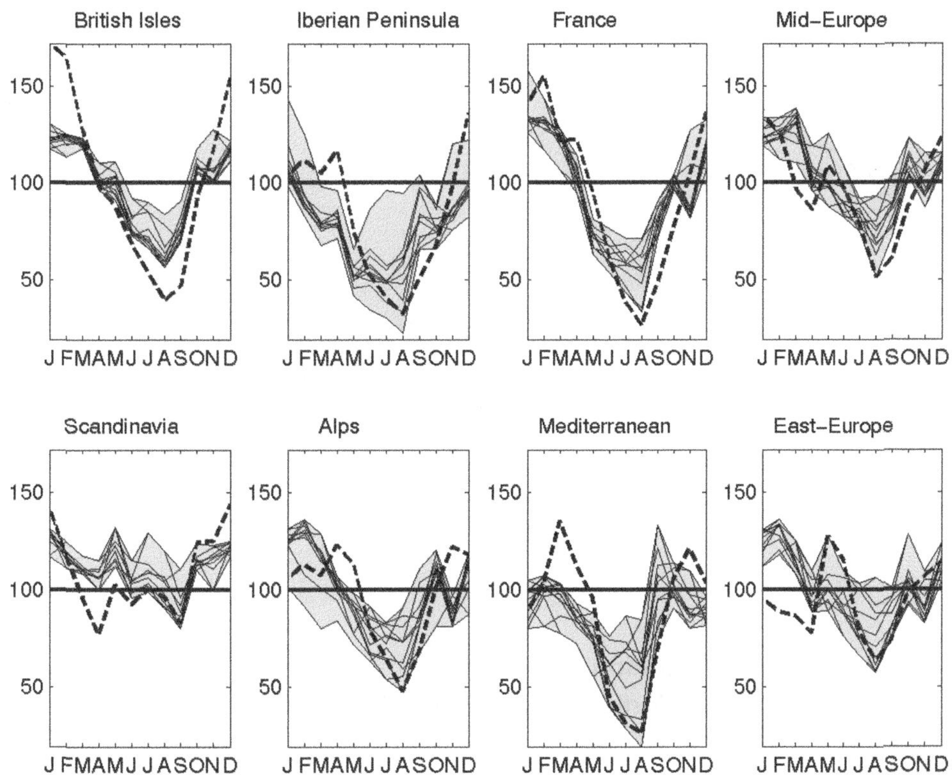

Figure 3: Monthly changes in precipitation for different regions in Europe simulated by an ensemble of different RCMs in the project PRUDENCE (Jacop *et al.* 2007). Shown is the difference 2071-2100 minus 1961-1990 for the A2 SRES scenario. The thick black line indicates the result of the driving GCM for the region.

7.3 OTHER DRIVING FORCES OF CHANGE

In addition to climate change, there are also pressures on water resources induced by human activities due to demographic trends, economic development, land use change, change in consumption patterns, and technological development. The pressures could be caused by a high population density, and/or an intensive land and water use. In some European regions human-induced impacts on water resources are currently more pronounced than the already observed climate change impacts (e.g. see Krysanova *et al.* 2006 and EEA 2006). The ATEAM project analysed the possible vulnerability of ecosystems and environmental services to climate *and* socio-economic changes. It is important to note that human interventions can intensify the pressures on water resources in the future by superimposing these on the climate change effects. However, there are also many examples, especially in terms of application of new technologies and sustainable land use management, where certain measures have the potential to increase the adaptive capacity of regions to climate change impacts and even to mitigate climate change. Impacts of different driving forces including socio-economic and demographic trends specifically on water resources and water availability were analysed in the project EEA OUTLOOK, financed by the European Environmental Agency.

7.3.1 Demographic trends

There are two main demographic trends, which can create notable impacts on local water resources: (a) population growth and (b) migration. The impacts of demographic trends on the water sector can be best seen in the African countries, where population growth is the largest worldwide (see Figure 4), and many people concomitantly flock to the rapidly growing cities. These new settlements often lack proper water infrastructure, and residents do not have sufficient access to clean water and sewerage systems.

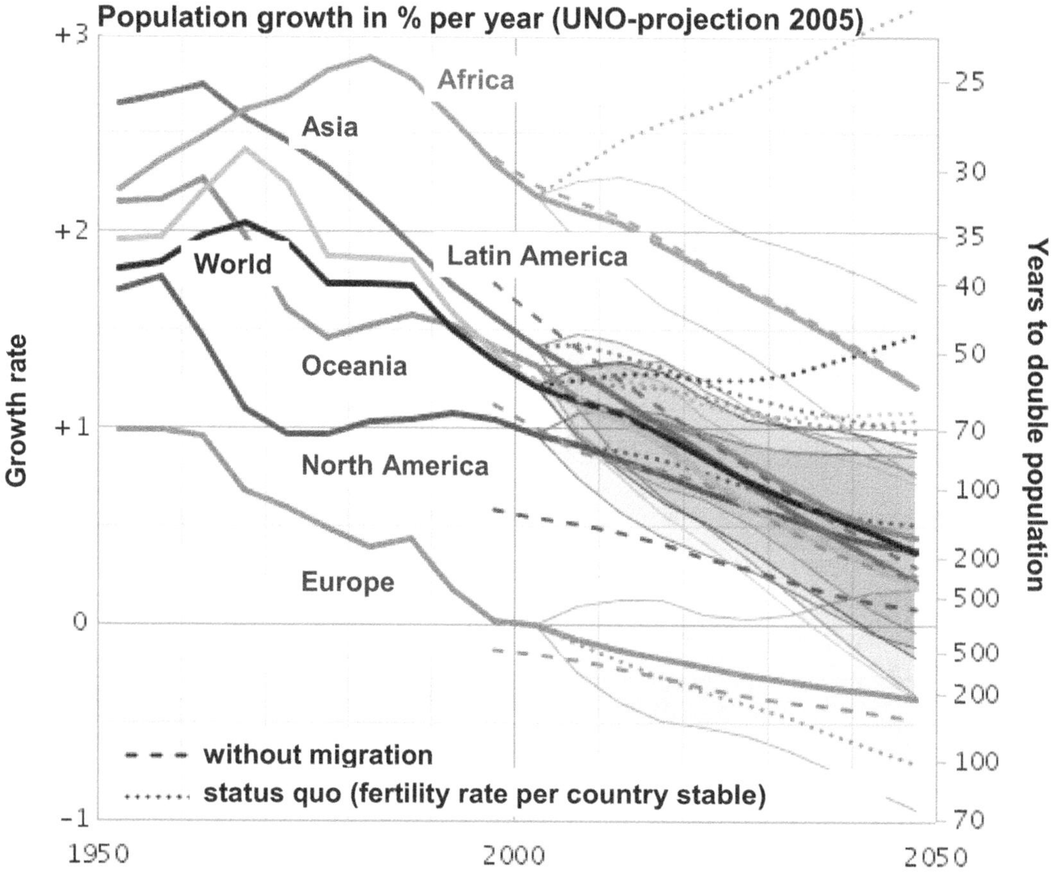

Figure 4: Observed and projected demographic trends (source: ttp://de.wikipedia.org/wiki/Bev%C3% B6lkerungsentwicklung, changed)

Since the 1950ies, the increase in world's population is slowing, and Europe is the first continent where the population started to decrease (Figure 4). The average birth rate in Europe is 1.5 children per woman (EUROSTAT). The countries of Eastern Europe undergoing political transition show a lower birth rate (Czech Republic 1.23, Poland 1.23, Russia 1.1), while France and Ireland have a nearly balanced population development (1.90 and 1.99 respectively).

Due to the population decrease and the relatively high technological standards, demographic pressures on water resources in Europe are mostly the result of permanent or temporary migration. A typical intra-national permanent migration is the depopulation of rural areas and the subsequent population growth of urban areas causing the so-called "urban sprawl syndrome". It is expected that the share of people living in urban areas in Europe will increase from approximately 75% in 2006 to more than 80% in 2020 (EEA 2006), attracting mostly younger people with higher education. A typical inter-national permanent migration is based on the search for well-paid employment in economically booming areas all over Europe. This trend is accelerated by the large gradient of payment from east to west and from south to north.

A typical temporal phenomenon is the seasonal migration of many million people to the attractive tourist destinations of the Mediterranean countries, the Alps, and attractive cities. In some European areas the number of tourists frequently outnumbers the local population by far. Quite often it happens in regions, where water resources are generally low, like on Mediterranean islands, and at times, where the water availability is low (in summer).

7.3.2 Land use and land cover change

Some changes in land use, like deforestation, expansion of arable land, drainage and melioration of wetlands and channelling of riverbeds caused very significant impacts on the hydrological systems over the last centuries. Originally, forest was the natural land cover nearly all over Europe, whereas nowadays agriculture dominates, especially in areas with soils of high fertility.

Three major land developments can be distinguished:

(1) Deforestation during the course of human settling and expansion of arable land following the Neolithic revolution,
(2) Drainage and melioration of wetlands and channelling of riverbeds in the 20th century aimed in gaining additional arable land and improving navigation, and
(3) Urban sprawl around large cities and industrial centre regions during the last decades.

The climax of *deforestation and expansion of arable land* occurred in the late Middle Ages, when, for example, the share of forest in Germany was only 12 % (compared to about 33 % nowadays) leading to the introduction of the first forest conservation laws in Europe. Forest has much higher interception (especially during winter) and deeper roots than grass and agriculture crops, and therefore much lower total runoff than arable land. Radiation images of European landscapes also show that the near-surface atmospheric temperature in sunny days is some degrees lower in forest than in arable land, resulting in different energy balances and hydrological patterns. Erosion process intensifies without forest cover, and as a result in large parts of the Mediterranean area entire landscapes lost their soil cover, so that flash floods cannot be buffered.

Most of European cities have been built along river courses, like e.g. Paris, London, Berlin and Cologne. This was only possible through the implementation of drainage systems to drain wetlands and the diversion of riverbeds. The same happened at large scale in rural areas with the aim to gain additional land of good quality for agricultural purposes. Especially extensive *drainage and melioration of wetlands, straightening of rivers and channelling of riverbeds* occurred in Europe and worldwide in the 19th and 20th centuries. For example, the Elbe River lost 86 % of its floodplains, and the length of the main river reach was reduced from 1.206 km to 1.091 km (ATV-DVWK 2000). However, these measures also had unforeseeable and undesirable consequences, like changing the hydrological pattern of runoff generation and thus river discharge characteristics due to decrease of the flood buffering capacity, and reduction of filtering capacity for pollution from upland areas.

Urbanisation changes the hydrological pattern of the landscape because the share of sealed areas is becoming higher while the natural river flow is regulated. Often the water demand of these conurbations is much higher than the local water availability, and water has to be withdrawn from the nearby rivers or to be brought from remote reservoirs. Possible impacts of urban sprawl are the generation of flash floods and the deterioration of water quality.

The most important land use changes taking place in Europe today are urban sprawl, changes in forest management, changes of the agricultural management practices like structural changes in the former socialistic eastern European countries, and reallocation of arable land for bio-fuel production. It is important to take into account that land use management also has a high potential to adapt to or even to mitigate climate change.

7.3.3 Consumption patterns and technology trends

While the technical development in Europe over the last centuries was mostly associated with an increasing water demand, the last decades showed an opposite trend: technical improvements, stimulated by high prices for drinking water and cleaning of wastewaters, in addition to a new system of laws for water protection, lead to a decrease of water use in many European countries. This also happened in the eastern part of Europe over the last decade, where the change from a planned economy to a more market-oriented one, closure of many industrial enterprises in connection with the improvement of the water infrastructure (modern sewage treatment plants, reservoirs, pipelines and channels) and higher water prices lead to a notable decrease of the industrial and private water use and to some increase of the water quality. The water use for irrigation of arable land, vine yards and vegetable-growing areas in Europe is the highest in the Mediterranean, where the trend to decreasing precipitation is the highest in Europe, and in some areas groundwater abstraction already exceeds the recharge of groundwater.

7.4 IMPACTS AND CONSEQUENCES FOR WATER RESOURCES AND WATER-RELATED SECTORS IN EUROPE

7.4.1 General impacts on water balance components

Change in climate will inevitably cause changes in the water balance components: evapotranspiration, river discharge and groundwater recharge, which determine the availability of water resources. To assess the impacts of climate change on the water balance components, a coupling of atmospheric and hydrological models is required. Here one of the main problems is the scale difference between the climate and hydrological systems. With GCMs having a quite coarse spatial resolution (grid size of several thousand km^2), hydrological models require much finer detail (a few km^2). The two main techniques to overcome this problem are the use of statistical downscaling methods or Regional Climate Models (RCMs). Changing water resources in Europe under climate and land use change was the topic of different European projects, e.g. of the projects EEA OUTLOOK, ATEAM and of the still ongoing project ENSEMBLES (see Table 1), applying different downscaling methods.

Generally, **evapotranspiration** demand increases due to climate warming. Evapotranspiration is influenced mainly by the availability of energy (solar radiation) and water, as well as by air temperature, air humidity, wind speed, vegetation and soil characteristics. However actual evapotranspiration rates are constrained by the actual water availability, and higher temperature could result in lower actual evapotranspiration if water availability decreases. Generally, in areas with sufficient available soil moisture (e.g. northern Europe) climate warming will lead to an increase in evapotranspiration. By contrast, in southern Europe (e.g. the Mediterranean region) a reduction in water availability during the summer season will lead to a reduction in total evapotranspiration despite a temperature driven increase in evaporative demand.

River discharge. Changes in annual river discharge are projected to vary significantly across Europe, related to regional environmental settings and local changes in precipitation and temperature. As a general trend demonstrated by many modelling studies, runoff in higher-latitude areas with increased precipitation will most likely increase in winter and decrease in spring (less precipitation as snow in winter and less snow melting). Summer flow in river basins with considerable groundwater contribution will change in accordance with changes in precipitation during the groundwater recharge period in winter. In arid and semi-arid Mediterranean areas change in precipitation will be translated into runoff change. In all cases change in runoff will be considerably higher in percentage than change in precipitation due to the non-linearity of response.

According to the results of WaterGAP model (Lehner *et al.* 2001 & 2005), only small changes in average annual river discharge (-5 to +5 %) can be expected in most of Europe by 2020, except some regions in the Mediterranean region and eastern Europe, where decreases up to 25 % are possible, and some regions in northern Europe with up to 25 % increases. By 2070s the same signals of changes are more explicit with river discharge decreasing by more than 25 % in southern and south-eastern Europe, and increasing by more than 25 % or more in most parts of northern and north-eastern Europe (Lehner *et al.,* 2001 & 2005). Changes in runoff regime and seasonality are also expected to occur under climate change. By 2020 a general increase in winter runoff and decrease in summer runoff can be stated. Decrease in summer runoff in southern Europe may be up to 50 % by 2050 and up to 80 % by 2080 (EEA 2005a).

However, predictions of climate change effects on river discharge are still quite uncertain and the predictions of different models are variable. The uncertainties are mainly caused by uncertainties in the projections of changes in precipitation, with superimposed uncertainties in hydrological models and their parametrisation.

Groundwater recharge. Even relatively small reduction in precipitation in conjunction with temperature rise may notably impact groundwater recharge, potentially leading to a recharge reduction (Kruger *et al.* 2001; Eitzinger *et al.* 2003; Arnell 2003; Hattermann *et al.* 2004). Rising temperature induces longer vegetation growth periods. This causes an increase in evapotranspiration, which continues later into the autumn season leading to delayed and lower groundwater recharge. Also, the end of the recharge period would come earlier in spring. This would be most evident in south-eastern Europe. Cooper *et al.* (1995) found that the effect of various climate scenarios on groundwater recharge depends on the aquifer type, and that scenarios incorporating higher evaporation resulted in the highest change in the hydrological regime of aquifers.

Besides, the coupling of climate change and anthropogenic activities could be even more important for groundwater. Aquifers usually recharge very slowly, and the current level of water abstraction may appear unsustainable under future climate conditions, especially in some southern European regions, where aquifers are already overexploited and falling groundwater levels are observed (Eisenreich 2005).

Water resources availability. The above stated changes in water balance components under climate change translate into changes in water resources availability. Apart from climate change effects on the water cycle, human-induced pressures (mainly in domestic, industrial, and agricultural sectors) control the availability of water resources in Europe. A large proportion of Europe's population already lives in watersheds with less than 1700 m^3 $capita^{-1}$ $year^{-1}$ (the water-stress threshold defined by the World Bank), and beyond the 2020s climate change would increase the number of people under this threshold by tens of millions. Information on the combined impact of climate and human water withdrawals can be also found in studies by Lehner *et al.* (2001) and Döll (2005), indicating increasing pressure on water resources in most river basins of southeastern, southern and parts of central Europe by 2050s compared to situation in the 1990s.

7.4.2 Impacts on extreme events (floods and droughts)

A cluster of projects deals with modelling of extremes under climate change, including STARDEX, PRUDENCE and MICE. A core element is to explore the uncertainty in climate projections by use of ensemble runs. The project MIND focuses especially on droughts in the Mediterranean, the project CICILIA on climate modelling and impacts on hydrology in central and eastern Europe.

Floods. According to the Clausius-Clapeyron law, the atmosphere's water holding capacity increases with temperature, and the potential for more intense precipitation increases. Higher and more intense precipitation has already been observed globally, and this trend is expected to increase under global warming. This is a sufficient condition for flood hazard to increase (Kundzewicz and Schellnhuber 2004). Besides, there are other, non-climatic factors, like land use change escalating flood hazard. Among them are deforestation, urbanisation, elimination of floodplains and wetlands, and construction activities in flood-prone and coastal areas.

Under the scenario of CO_2 doubling, climate models are converging towards an intensification of the hydrologic cycle with an increase in precipitation intensity, up to 1 % per decade in the 21th century over the northern mid- and high latitudes, whereas other areas in southern mid-latitudes will be experiencing drier conditions (IPCC 2007a). The results of climate models have also shown that the risk of extreme precipitation and flooding would increase in the future (Milly *et al.* 2002; Palmer and Raisanen 2002).

Palmer and Raisanen (2002) analysed differences between the control model run with 20th century levels of CO_2 and an ensemble with transient increase in CO_2 around the time when CO_2 will be doubled (about 61-80 years from present). They found a considerable increase of the risk of a very wet winter in Europe. Their results indicate that the probability of total winter precipitation in the boreal zone to exceed two standard deviations above normal will increase by a factor of five to seven over large areas in Europe (e.g. in Scotland, Ireland, parts of Russia, and much of the Baltic Sea basin area).

Droughts. The latest climate impact studies suggest significant summer drying across many parts of Europe, particularly in the Mediterranean basin with more hot days and heat waves becoming very likely (Goodess 2005; IPCC 2007b; Lehner *et al.* 2005). A climate change impact study by Arnell (1999) has projected that droughts are likely to increase in frequency and intensity across most of western Europe under a range of climate scenarios. Simulations by Schär *et al.* (2004) suggest that every second summer in Europe by the end of 21st century would be as hot and dry as the summer of 2003. The results of these studies lead to the conclusion that over the next 100 years Europe is likely to experience more frequent meteorological drought conditions, especially in the southern regions. These events might manifest themselves either as short but extreme single season droughts (such as the hot summer of 2003) or longer-term, multi-seasonal droughts.

7.4.3 Uncertainties in research on extreme events

It should be clearly recognised that there is a high uncertainty associated with climate model results, impact assessment using climate scenarios, and estimation of future trends. Due to the complexity and nonlinearity of the interactions between atmosphere and the terrestrial water cycle, acting at various spatial and seasonal scales, some unexpected impacts could happen and some statements about impacts will inevitably undergo correction in future.

For example, Schnur *et al.* (2002) stated that current climate models are not reliable at predicting extreme climate events in local areas, such as flooding in a given river basin, because they are limited in their resolution to a coarse grid size of about 200 km. For the average river basin, climate change simulations would need a much higher resolution of tens of km, but this will not be available in a near future. However, even the higher resolution of climate models *will not reduce uncertainty in model predictions of extreme events to zero* due to other sources of uncertainty, in particular, uncertainty related to GCM performance in simulating precipitation.

Until the high-resolution climate change simulations are available, downscaling methods need to be applied to produce the proper spatial resolution needed for extreme events modelling. Available downscaling methods are Regional Climate Models (using the GCM's as boundary conditions), statistical downscaling methods and stochastic downscaling. No clear consensus exists on which methodology is most reliable. Some authors, e.g. Arnell (1999) are sceptical and advise not to use downscaling at all.

7.4.4 Sensitivity of water-related sectors to climate change impacts

Reliability of water supply of adequate quality is one of the basic requirements of any human society and also a prerequisite for industrial production, agriculture etc. Most human activities rely on water in one way or other, and many of the key sectors of the European economies are sensitive to changes in the availability of water resources and extreme events. The sensitivity varies widely between sectors and regions. In the following, five water-related sectors having possibly the highest sensitivity to climate change in Europe (energy production, water supply and wastewater disposal, agriculture, navigation, and tourism) are discussed in more detail.

Energy sector is one of the key sectors of the European economies. Climate change itself will most likely have influence on both the demand and supply of energy. Energy use might change as a direct effect of temperature changes. As a result, the overall energy consumption for heating would decrease and that for cooling would increase (both in terms of overall energy supplied, also to meet peak demands). Therefore, benefits from increased winter temperatures that reduce heating needs may be offset by increases in demand for summer air conditioning, as average summer temperatures rise (see European Commission (2005)).

A decrease in water availability may also be detrimental to energy and electricity production. Lower river discharge would reduce the hydropower potential and thus negatively impact on this mode of energy production. Thermal power stations, on the other hand, depend on large volumes of water for cooling. Severe water shortages, but also increased temperatures of cooling water, can thus lead to less efficient energy production or the temporary closure of coal-fired or nuclear power plants.

A new energy source promoted and subsidised by the EU is biomass production. However, the cultivation of crops for energy production is potentially subject to similar impacts described below for crop farming. Especially droughts can affect biomass plantations and consequently affect regional energy supply.

Infrastructure and services for **water supply and wastewater disposal** are directly affected by any changes in water resources availability. Increased precipitation and runoff will lead to higher water availability, but also intensify the pressure on drainage systems. In regions, where the frequency or duration of intense rainfalls or flooding events will increase, water quality problems associated with diffuse-source pollution (Krysanova *et al.* 2005), inundation of sewerage systems, or overflow from wastewater treatment sites would be aggravated. In some areas, however, an increase in water availability could be beneficial for water supply.

Water supply services in regions suffering from seasonal water stress (less precipitation, higher evapotranspiration or both) will be faced by the challenge to satisfy consumer demands despite water shortages. This problem could be intensified under drier conditions caused by climate change: lower runoff can lead to water quality problems, if the load of nutrients and other pollutants from treatment plants remains the same and is not adapted to the new discharge levels (point sources). In addition, higher temperatures support the growth of algae, thus increasing eutrophication and oxygen concentrations in surface waters. Low runoff levels would further accelerate this problem, leading to lower flow rates and less mixing of the water body in streams. This can again improve growing conditions for algae and decrease oxygen dilution.

Deteriorating water quality, which could result from saltwater intrusion into freshwater bodies and aquifers in coastal zones, might also pose a problem to water supply.

Agriculture, especially in arid regions, will suffer from the decrease in water availability caused by reduced precipitation or increased evapotranspiration levels (or both). Not only cash crops, but also live stocks, could be affected, especially during droughts or heat waves. The demand for irrigation is likely to increase because of higher potential evapotranspiration under warmer conditions. This might translate into significant additional costs for farmers, either from increased expenses for water pumping and transportation or from production losses.

However, in regions where higher temperature and/or precipitation levels reduce climatic constraints for agricultural production, the anticipated changes in climate may also turn out beneficial (Eisenreich 2005). In the northern latitudes of Europe, the agricultural sector may benefit from longer vegetation periods and higher crop yields, e.g. if farmers change from summer to winter crops. While new land may become available for arable farming in these regions, land may also be lost for farming in other areas, e.g. in coastal zones due to sea level rise, or in water-stressed regions due to desertification and salinisation processes.

Agricultural production in flood-prone areas could be affected by increased damages from flood events. Higher precipitation rates in combination with higher temperatures can also lead to production losses due to an increase of certain plant diseases, for example by providing better growing conditions for fungi.

Navigation on inland waters depends on certain ranges of water levels and is therefore sensitive to fluctuations in river discharge (floods and droughts). In this context, restrictions resulting from low-flow periods are more relevant for river navigation, since drought periods prevail usually longer than floods. A decrease in the base flow, resulting from melting glaciers, decreasing snow cover and higher evapotranspiration rates in a warmer climate, is projected for many European basins. This makes navigation less reliable, especially at the end of summer and early autumn.

However, river transport can also be disrupted during flooding events, and extreme floods have the potential to destroy harbour infrastructure and vessels as such. The same applies to the road or railway infrastructure. In regions, where an increase in precipitation is projected, low-flow periods may occur less frequent, which could have a positive effect for waterway transport. Still, sea level rise may still affect coastal harbours.

Tourism. Reliable climate conditions, like sufficient snow in winter sport resorts or warm and dry weather in summer tourist areas, is of a high importance for the tourism industry. A possible shift in precipitation and temperature patterns could thus lead to seasonal and regional shifts in tourist flows, causing a regional redistribution of income from tourism to a certain extent. Summer tourist flows to the Mediterranean might be reduced if heat waves, droughts and water shortages were to increase in frequency and intensity. In return, northern European and Alpine regions might become more attractive for tourism during summer, while loosing attractiveness due to snow cover changes (less snow/shorter period) in winter.

In addition, more indirect links between tourism and water-related impacts from climate change can play a role. Nature and wildlife, especially rare species, also constitute tourist attractions that might become threatened by climate change impacts. The relationship between tourism and climate change is particularly close in coastal areas, which constitute some of the most essential tourist resources. Here, sea level rise is putting recreational areas and popular tourist destinations (beaches, wetlands, estuaries and islands) at risk, while also endangering famous sights - in the extreme case of Venice a whole city (EEA 2005b p.29).

Although impacts on the following sectors are not discussed specifically, it should be noted that these sectors could also be affected by climate-driven changes in water resources. **Construction, building development and spatial planning** will face the challenge to prevent flood damage in river basins and coastal areas, and to protect settlements and infrastructure under more uncertain conditions and potentially under increased flood risk. Especially with regard to spatial planning, the importance of the natural environment, for example in providing water resources of sufficient quantity and quality and supporting biodiversity, should not be overlooked. This becomes particularly interesting in the context of conservation efforts of endangered habitats such as wetlands. The question of future conservation strategies, like aiming to maintain the status quo or following new favourable conditions needs to be discussed. Even, if the outcomes demand a change of previous land-use rights.

The forestry sector could be affected by similar processes like agriculture. **Fisheries**, on the other hand, depend on water temperature, quantity and quality conditions that are favourable to the sustenance of stable and healthy fish populations. While the **insurance** sector may be faced by an increase in insurance claims from flood and drought events, it might also benefit from a higher number of insurance contracts and an increased number of customers.

Overall, the considerable sensitivity of different socio-economic sectors to climate-driven changes in water resources is evident and a wide spectrum of impacts, both positive and negative, can be expected. Ultimately, the net effects will be highly dependent on regional conditions and the adaptive capacity of each specific sector.

7.5 COPING STRATEGIES NOW AND NEEDED IN FUTURE – A EUROPEAN POLICY PERSPECTIVE

The previous sections discussed climate change impacts on water resources and water-related sectors in Europe. However, vulnerability of the water-related sectors to climate change depends not only on their *sensitivity* to climate change as such, but also on their *exposure* to climate change and on their *adaptive capacity* (Schröter *et al.* 2004). Infrastructure is not affected by a flood if not located in the flooded areas, and the vulnerability of the transport sector can be reduced by fostering a mix of different transport modes to increase the adaptive capacity. Adapting to climate change impacts thus means reducing the vulnerability of societies to these impacts, either by limiting the exposure, or by reducing the sensitivity of the sector to climate change, or by increasing its adaptive capacity.

Adaptation to climate change is possible at different levels: (1) at the technical level (e.g. by applying flood protection measures such as reservoirs, dams and dykes), (2) at the management level (e.g. developing intelligent management schemes to save water during periods of high water supply to cover periods of low water supply), and (3) at the policy making level (e.g. establishing suitable legislations to protect water resources and to influence the amount of water consumption by taxes or subsidies).

Adaptation to water-related climate change impacts is gaining relevance both on the political and the research agenda in Europe. The European Environment Agency (EEA) has been pushing water adaptation research in recent years (EEA 2005b; 2007). In September 2006, the European Commission (DG Research, DG Environment, and Joint Research Centre) organised a workshop in Brussels, which discussed climate change impacts on the water cycle and promoted dialogue between scientists and policy-makers.[1] In February 2007, the German EU presidency in co-operation with the European Commission hosted an international symposium in Berlin, which discussed impacts of climate change on water management and water dependent sectors as well as options for adaptation.[2] In order to inform this conference, the German government funded a research project on climate change, water and adaptation, which was carried out by Ecologic and the Potsdam-Institute for Climate Impact Research (Leipprand *et al.* forthcoming). The European Commission launched a Green Paper on Climate Change and Adaptation in July 2007.

7.5.1 Technical measures and management strategies to cope with climate change

Combinations of technical measures and improved management represent an appropriate strategy to adapt to climate change, because they can be implemented within a single sector and at the local or regional scale. Although they are often meant to decrease the vulnerability to climate change of a single sector or region, they most often affect also the vulnerability of other sectors or regions. For example, measures to increase the reliability of navigation during low flow periods could also increase the reliability of power production in case a major part of the fuel is transported by vessels. Building of reservoirs will influence the hydrograph of the entire river, also in the downstream areas. Therefore an integrated approach to water resources management (IWRM), especially if it involves relevant stakeholders in the decision making process, is so important to provide a sustainable and widely accepted management solution (see also Section 6).

Technical measures and new management strategies can be stimulated and supported by political decisions, but do not necessarily depend on them. Table 3 summarises major technical measures and management strategies to adapt to climate change impacts in terms of floods, droughts and sea level rise. Many more measures exist (see e.g. Kabat *et al.* 2004). The appropriateness of these measures should be considered in the regional context, and they can be implemented taking the local characteristics of the natural and social environment in the region into account.

Table 3: Technical measures and management strategies to adapt to climate change impacts on hydrology

Flood protection
Technical flood protection (e.g. raise dykes, enlarge reservoirs, upgrade drainage systems etc.)
Natural retention of flood water (e.g. floodplain restoration, change of land use)
Restriction of settlement/building development in risk areas
Standards for building development (e.g. permeable surfaces, greening roofs etc.)
Improving forecasting and information
Improving insurance schemes against flood damage
Drought/low flow protection
Technical measures to increase supply (e.g. reservoir volumes, water transfers, desalinisation etc.)
Increasing efficiency of water use (e.g. leakage reduction, use of grey water, more efficient irrigation etc.)
Economic incentives (e.g. water pricing)
Restriction of water uses
Landscape planning measures to improve water balance (e.g. change of land use, reforestation, reduced sealing of areas)
Improving forecasting, monitoring, information
Improving insurance schemes against drought damage
Coastal zones
Reinforce or heighten existing coastal protection infrastructure
Retreat strategies, e.g. managed realignment of dams
General adaptation measures
Awareness raising, information campaigns
Building of financial resources

1 http://ec.europa.eu/research/environment/newsanddoc/other_pubs_en.htm.

2 A number of background documents, presentations and the conclusions are available from the conference website at http://www.climate-water-adaptation-berlin2007.org/.

A survey conducted in 2006 among European water managers (EEA 2007; Leipprand *et al.* forthcoming) showed that there is high awareness about climate change effects on the water sector, and that European countries expect significant changes in water resources and hydrology. However, efforts to implement measures and initiatives to adapt to water scarcity and drought seem to be lagging behind. Individual projects are being implemented in the European Union, although many are still at the planning stage. Few Member States have developed comprehensive adaptation strategies to date, and few governments seem to move on to implementing adaptation initiatives and incorporating long-term climate change risks into actual investment or development plans on a national and local scale (see also EEA 2005b; Gagnon-Lebrun and Agrawala 2006). Generally, adaptation activities related to water currently seem to be focused on flood management and defence, while adaptation measures related to the management of water scarcity and drought, although recognised as equally important, do not yet seem to be widespread.

7.5.2 Role of EU policies in supporting adaptation

While adaptation projects on the ground will be implemented mainly in a decentralised manner and by individual private actors or local or regional authorities, there is a broad consensus in the scientific literature that policy at higher levels may play an important role in supporting adaptation. Government action can contribute to motivating early adaptation and to support efforts by private actors, to funding for research, improving communication and dissemination of research results, and to integrating climate change concerns and adaptation into sectoral and planning. European water managers have expressed an interest in support and co-ordination of their adaptation efforts at EU level (EEA 2007; Leipprand *et al.* forthcoming).

Below is a brief overview of the European policies that are relevant in the context of adaptation to climate change is given, and the scope for integrating adaptation to water-related climate change effects into these policies is explored.

The Environmental Action Programme

The main pillars of the EU environmental activities are formulated in the Sixth Environmental Action Program (EAP) entitled "Environment 2010: Our Future, Our Choice", with one of the four action points of the program being climate change and global warming. In addition, the Action Programme emphasises the importance of taking the environmental impact into account in all relevant EU policies (e.g. agriculture, development, energy, fisheries, industry, the internal market, transport). With regard to climate change, the 6th EAP suggests that the Community should prepare measures aimed at adaptation to the consequences of climate change by:

- Reviewing Community policies, in particular those relevant to climate change so that adaptation is addressed adequately in investment decisions; and
- Encouraging regional climate modelling and assessments both to prepare regional adaptation measures such as water resources management, conservation of biodiversity, desertification and flooding prevention and to support awareness raising among citizens and business.

The European Climate Change Programme

In 2000, the EU launched the first European Climate Change Programme (ECCP), a consultation process which aimed to identify and develop the necessary elements of an EU strategy to implement the Kyoto Protocol and involved a multitude of stakeholders. While the first ECCP focussed on mitigation and prevention of climate change, the second ECCP, launched in 2005, also recognised adaptation as an important issue and created additional working groups to discuss adaptation issues. Efforts to mitigate climate change are also continued and further developed under the ECCP II. The first policy outcome of the Commission's ECCP activities on adaptation is the Green Paper launched in July 2007.

The Water Policy

The adaptation of water management to climate change requires planning under uncertain conditions and a constant consideration of the potential risks and the costs and benefits connected with adaptation measures. The main aim of the current European Water Policy is to clean up polluted waters, and to ensure that clean waters are kept clean and available in a sufficient quantity. Up to now, climate change impacts have not been a major concern in EU water policy, and accordingly the water legislation in its current state contains only a limited number of explicit references to climate change impacts. However, European water policy is characterised by a broad scope, and it provides a range of instruments to address a number of different problems. It therefore provides many opportunities to integrate and support adaptation efforts. The EU Water Policy is realised through Directives, framework legislation that has to be implemented in the Member States with varying degrees of flexibility. The three Directives most relevant in an adaptation context are the Water Framework Directive, and the forthcoming Marine Strategy Directive and Flood Directive.

While earlier EU water legislation addressed different problems individually, the *Water Framework Directive* (WFD), which was adopted in 2000, introduces integrated water management at river basin level as a new and innovative approach. The main objective of the Directive is to ensure that by 2015 all European waters achieve "good status" – a concept which integrates chemical, ecological and morphological indicators. For each river basin, pressures on water have to be identified, and River Basin Management Plans (RBMPs) have to be designed that introduce measures to work against these pressures and achieve the "good status" objective. These RBMPs have to be revised periodically.

The WFD does not explicitly make reference to climate change. However, it addresses several issues that are relevant in a climate change context. For instance, the mitigation of floods and droughts, which are likely to become more frequent in a changing climate, is mentioned as one of the main purposes of the Directive. Furthermore, climate change impacts may exert pressures on water bodies and thus interfere with several of the key phases and with the delivery of key environmental objectives of the WFD (Eisenreich 2005; Wilby *et al.* 2006). On the other hand, if climate change impacts and adaptation concerns are considered in the WFD implementation process and in river basin management planning, it may provide a powerful tool for supporting climate change adaptation policies on a national and regional scale. The WFD's integrative approach to management furthermore allows encouraging the contribution to adaptation efforts by all relevant sectors and stakeholders. In the context of the Water Framework Directive implementation process, a European working group was set up in 2007 that will explore potential adaptation strategies in the context of the European water policy.

Marine Strategy Directive. For the protection of the marine environment, the European Commission has proposed a Marine Strategy Directive that in many aspects follows a similar approach as the WFD. The central aim is to achieve good environmental status of the marine waters by 2021 and to protect the resource base upon which marine-related economic and social activities depend. In a similar manner as with the WFD, the Member States have to identify pressures and respond to them by defining specific measures. The Marine Strategy Directive identifies climate change as one of the main pressures on marine waters, alongside the impacts of commercial fishing, loss or degradation of biodiversity, and contamination. The protection of the marine environment should be flexible enough to allow for an adaptation to changing pressures and impacts, which may for instance be caused by climate change. While it does not suggest any specific adaptation measures, the Marine Strategy provides a framework and policy instruments that can be used to promote adaptation.

The proposed Marine Strategy Directive is part of a European Maritime Policy that is currently being developed. This Policy also recognises that adaptation strategies including the organisation of sea defense will be required to manage risks for coastal and offshore infrastructure resulting from sea level rise, increased flooding and storm surges. It also mentions that Mediterranean coastal zones are likely to be affected by changing precipitation patterns, and that an increased need for desalinisation may result from this.

Flood Directive. In January 2006, the Commission proposed a Directive on flood risk management, after extensive discussions with relevant stakeholders and the public. In April 2007 agreement was reached between the European Parliament and the Council on the final text of the Directive; formal adoption by the Council is expected later in 2007. The Directive introduces a three-step approach. First, Member States have to undertake a preliminary assessment of flood risk in river basins and coastal zones. Where significant risk is identified, flood risk maps have to be developed. Finally, flood risk management plans must be developed for these zones. These plans have to include measures that will reduce the probability of flooding and flood damage, and that should focus on prevention, protection and preparedness.

The rationale of the proposal recognises that climate change, together with inappropriate river management and construction in flood risk areas, might increase the scale and frequency of floods in the future. The Directive text itself also makes reference to climate change, stipulating that projected climate change should be taken into account in the assessment of future flood risk and of its consequences on human health, the environment, and the economy. The risk management instruments introduced by the Directive will represent powerful tools for addressing increased flood risk caused by climate change. In combination with the WFD, the Floods Directive would allow taking full account of climate change effects on flood risk, and would encourage water managers to respond to challenges by means of an integrated management approach..

The Sectoral Policies (agriculture, energy)

Energy production. Most of the energy production in Europe is connected to the use of water (cooling, hydropower, growing of biomass). Besides, it should be also kept in mind that energy production is one of the major drivers of climate change. This means that the EU Energy Policy has to consider both sides, measures to mitigate and prevent climate change and measures to adapt to climate change in order to secure energy production. In fact, while the EU is taking steps to address its greenhouse gas emissions in the energy sector, adaptation of the energy sector to climate change currently does not have a high priority on the political agenda. Nevertheless there are several elements in the current policy framework that can provide starting points for the development of adaptation strategies.

The current strategic approach of the EU to energy policy (Green Paper on developing a common, coherent European Energy Policy: European Commission, 2006) does not address climate change impacts and adaptation. However, it includes elements that are relevant with respect to adaptation. For instance, it envisions a review of the EU's Energy Policy and action to improve security of supply. In this context, resilience to climate change impacts of different energy sources or technologies might be a relevant criterion.

Agriculture. The Common Agricultural Policy (CAP) provides financial support to farmers from two different funds. The majority of subsidies are direct payments to farmers and product-related subsidies. The second, smaller fund supports measures intended to strengthen rural development.

Both of these pillars of the CAP influence the vulnerability of European agriculture to climate change impacts. Payments under the first pillar used to be coupled to the quantity of agricultural production and thus in many cases provided incentives for unsustainable and non-adaptive farming practices, such as the growing of irrigation-intensive crops in dry regions. Recent CAP reforms have reduced these incentives by de-coupling payments from production.

The Rural Development policy, the second pillar of the CAP, also provides opportunities to support adaptation. Both mitigation of climate change and adaptation to climate change impacts are acknowledged as priorities in rural development (RD) funding, and Member States are encouraged to implement appropriate actions. RD programmes may for instance provide support to farmers for purchasing new equipment needed for adaptation, for the development of new products, processes and technologies, or for educational measures.

Science and Technology Policy

Adaptation to water-related climate change effects will require further research to project the impacts at regional level and to facilitate early prediction of natural disasters in order to enable local and regional public and private sector actors to develop cost-effective adaptation options. Further, adaptation to climate change will also require technological innovations and investment. Two instruments set up by the EU are relevant for addressing these needs: The Environmental Technology Action Plan (ETAP) aims to reduce pressures on our natural resources, improve the quality of life of European citizens and stimulate economic growth. The main instrument for funding research and development in Europe is the Framework Program for Research. The new programme for the period 2007 - 2013 (FP7) organises collaborative research into nine themes related to the environment, including the issue of climate change. Both mitigation and adaptation play an important role.

7.6 SUMMARY AND DISCUSSION

Global observations show that the climate of the earth undergoes a significant change, and the global warning is an established fact (IPCC 2007a). Following the law of Clausius-Clapeyron this will lead to an increase of evapotranspiration and hence to a higher water holding capacity of the atmosphere and to an intensification of the water cycle.

However, the regional impacts of climate change may differ over the European continent because of regional specific circulation patterns and topography. Although climate change research is well established, it is currently difficult to project with certainty all consequences of climate change for a certain river basin, country or region. Relatively certain is the trend in temperature of 1.0 to 5.5 degrees over the next century. The uncertainty range is due to different scenario assumptions and model uncertainty.

More uncertain is the change in precipitation over Europe, although the observed general trends and the projections for the future show an increase in precipitation in northeast and parts of western Europe and a decrease in parts of central Europe and the Mediterranean. This general trend is overlaid by regional characteristics and seasonal trends: although the annual trend in a certain region shows a decrease, it might be that the winter precipitation would still increase. These seasonal effects are important because they could be more extreme than the general trend (e.g. summer droughts). Higher and more intense precipitation has already been observed globally, and this trend is expected to increase under global warming, which is a sufficient condition for flood hazard to increase (Kundzewicz and Schellnhuber 2004).

The consequences for the water-related sectors are manifold. Water-related sectors in many parts of Europe, especially in the south, will have to deal with lower water supply under warmer and drier climate. The central and northern parts of Europe will have to cope with lower water supply in summer. This could lead to disruption of navigation, impacts on vegetation (crops) and problems for providing sufficient cooling water at power plants. Increase of water supply and temperatures in northern and western Europe can be beneficial for water-related sectors, especially for agriculture (longer vegetation period) and hydropower plants (more water). However, the increased water supply may lead to a situation when the limit loads for drainage systems and treatment plants are exceeded, and the warmer and more humid climate can also stimulate development of crop diseases. Besides agriculture, tourism may be the next sector most sensitive to climate change, in particular in terms of winter sport, where the climate change impact on snow cover challenges winter sport resorts located at lower altitudes. A negative effect can be that the tourism flow pattern as such may change when northern locations would become more attractive during summer under warmer climate.

Summarising the discussion, consequences of climate change for water-related sectors could be severe. On the other hand, these sectors had to deal with climate variability in the past and this experience could help to adapt also to climate change (Kabat *et al.* 2002). However, it is also likely that the ongoing climate change will lead to situations or extremes never faced before, and new technical measures or management strategies or combinations of both have to be implemented. Here, experiences gained somewhere else under comparable situations may help to adapt, and the EU should support adaptation measures by its legislation (see Section 5).

One major problem is that the trend in climate is not steady but overlaid by climate variability. It is therefore uncertain whether a climate extreme is a climate change signal, and this uncertainty may lead to a situation where a single actor (farmer, company) may invest in inadequate adaptation measures, which could be either inadequate in space (regional impact insufficiently addressed) or inadequate in time (too early and investments do not pay off, or too late and damages are irreversible), although the sector as such will adapt to climate change over time. Therefore adaptive management as a systematic process for continually improving regional management policies and practices by learning from the outcomes of implemented management strategies is of a high priority, and a further investigation of

- Regional climate change impacts and possible adaptation measures,
- Adaptive management flexible enough to take into account new developments, and
- Support of national and European legislation for adaptation is important.

In February 2007, an international symposium (called "Time to Adapt - Climate Change and the European Water Dimension") was hosted in Berlin by the Federal Ministry for Environment, Nature Conservation and Nuclear Safety as part of the German EU presidency's activities. Representatives from governments, science and research, stakeholder groups and non-governmental organisations discussed the likely impacts of climate change on water management and water dependent sectors such as agriculture, energy, inland navigation and tourism, as well as options for adaptation. Aim of the symposium was to identify possible adaptation strategies in water related sectors and knowledge gaps and to strengthen the political profile of these issues. The results of the Symposium were summarised in a set of conclusions and recommendations (see http://www.climate-water-adaptation-berlin2007.org/). Preparation of the international Symposium involved consultation with scientific groups and major stakeholders, as well as close cooperation with the European Commission. An international Steering Group was set up to advise the convenors in all organisational and content-related matters in the course of conference preparations. The conclusions of the symposium in terms of knowledge gaps and recommendations for further research can therefore be seen as a comprehensive summary representing the current burning issues in climate change research. The key issues read as follows:

A. Better understanding and quantification of uncertainty throughout the chain of "emissions → climate change→ physical impact → ecological impact → socio-economic impact and improving the communication and handling of uncertainty in political decision making processes.
B. Better understanding and quantification of economical and social impacts of climate change in the different sectors.
C. Further improving and linking climate, hydrological, bio-physical and socio-economic models to better understand the complexity of the water cycle and aquatic ecosystems and how these will react to climate change. Since projections of changes at regional or river basin level will be most relevant for adaptation, the downscaling of models to lower scales needs to be a focus of future research.
D. Identifying thresholds and points of no return beyond which recuperation of the water resources and the water dependent systems is no longer possible.
E. Options for adaptation strategies, which integrate sectoral and cross-sectoral measures, and the assessment of their ecological, social and economic potential, benefits and costs.
F. Interdisciplinary approaches will be key, and attention should be given to the design of organisational structures that increase the capacity of water management to adapt to climate change.
G. Establishing of a solid baseline for evaluating Member State's adaptation plans, based on a historical archive of climatology.

From the list above it becomes clear that only the integration of the relevant economical sectors and also the natural environment can lead to sustainable solutions avoiding undesirable impacts on water resources. This is especially important considering the inherent uncertainty in projected climate change. What is therefore needed is an interdisciplinary approach flexible enough to allow changes in water management planes to adapt to a variety of climate change impacts in future, if possible in a proactive way. This is way Integrated Water Resources Management (IWRM) and Adaptive Management should play a key role in water resources management.

7.6.1 Role of Integrated Water Resources Management

In most parts of the world, the hydrological character of catchments, their vegetation and soil patterns, status of water bodies and flow regime have been significantly changed by humans and their region-specific water and land use management. Any observed trend is most often the result of a variety of many interacting processes, measures and interventions, with climate change or climate variability constituting only one of them. The complexity of the problem makes it necessary to consider all different land and water users in a river basin applying an integrated approach, which enables to take into account all major dependencies and feedbacks (Bronstert *et al.* 2005). This is the core of the Integrated Water Resources Management (IWRM) concept, which attempts to consider the catchments and their different users as a whole. It assumes that most interventions in the water cycle will very likely affect all, especially downstream users (see also Chapter 1).

According to van Beek *et al.* (2002), IWRM is the encompassing paradigm for adaptation to contemporary climate variability and it is the prerequisite for coping with the still uncertain consequences of global warming and their repercussions on the water cycle. The recent understanding of IWRM has been defined by the Technical Advisory Committee of the Global Water Partnership as follows: "*IWRM is a process, which promotes the coordinated development and management of water, land and related resources, in order to maximise the resultant economic and social welfare in an equitable manner without compromising the sustainability of vital ecosystems*" (TAC 2000).

One of the earliest attempts to define an appropriate methodology for integrated river basin management is the so-called DPSIR scheme of Driving forces - Pressures - States - Impacts - Responses suggested by the European Environment Agency (OECD 1994; UNCSD 1996). The scheme describes a natural system subject to anthropogenic pressure, where the Driving forces (e.g. land use, climate) generate a Pressure (e.g. climate change, land use change) upon the system (the river basin), thus altering the State of the system (water availability and quality). This alteration represents an Impact, i.e., an effect upon the environment and society, for example a decrease in water availability or quality. When this affects society in an unfavourable way, society reacts by devising and implementing Responses (e.g. adapted water and land use management) that can target either the Drivers, the Pressures or the State so that undesired or threatening Impacts are avoided or compensated (Soncini-Sessa 2004).

The concept of IWRM is also one of the key elements of the European Water Framework Directive (WFD), which demands to bring the surface water bodies in Europe into a "good ecological status" and the groundwater bodies into a "good chemical status". To support the implementation of the WFD, several projects have been launched, aiming at different issues like model-support, definition of quality standards and stakeholder participation. The EU concerted action Harmoni-CA has the objective to collect the information available in Europe and to bridge between science and practise. A web portal is currently developed combing and presenting the relevant projects and experiences (www.wise-rtd.info).

7.6.2 Role of Adaptive Management

A problem for IWRM is that the complex nature of management problems in river basins makes it practically impossible to design for example an ultimate River Basin Management Plan (RBMP) according to the Water Framework Directive already at the beginning of the implementation process. It is rather likely that unforeseen problems will occur or that new pressures will make it necessary to adjust the planned program of measures and it will be hence necessary to adapt the RBMP to these new pressures or problems. One of such potential problems is climate change and its impacts, which is associated with high uncertainty. It is therefore crucial that the planning framework is flexible enough to adapt to new problems and solutions. A broad participation process including all relevant stakeholders could help to identify all possible alternative management strategies and also to enable better implementation of measures later on. This is the core of the Adaptive Management concept. However, such participation process is usually not easy to organise, and the search for an optimal management strategy could be also a problem, if different stakeholder groups have different priorities (see also Chapter 2). The aim of the EU project HarmoniCOP was therefore to increase the understanding of participatory river basin management planning (RBMP) in Europe. The major outcome of the project is a handbook ("Learning together to manage together – Improving participation in water management"), presenting innovative support for the follow up of social learning in participatory processes of water management. New approaches to adaptive water management under uncertainty are the research topic of the EU funded project NeWater, identifying key elements of current water management regimes and their interdependence and promoting a transition to new approaches to adaptive water management.

According to Pahl-Wostl (2005), the major aim of adaptive water resources management is to increase the adaptive capacity of the water system or river basin. Adaptive Management is thereby a systematic process for continually improving management policies and practices by learning from the outcomes of already implemented management strategies (lessons learned). The most effective form of Adaptive Management employs management programs that are designed to experimentally compare selected policies or practices, by evaluating alternative hypotheses about the system being managed. It is aimed at integrated system design. The problem to be tackled is to increase the ability of the whole human-technology-environment system to respond to change rather than reacting to undesirable impacts of change. Hence it is a pro-active management style. Increasing the ability for change includes for example increasing the use of small-scale technological solutions or combining formal regulations with informal institutional settings.

7.7 CONCLUSIONS

Climate change research is well established in Europe. Over the last two decades, several projects funded by the EU investigated climate trends and their impacts on water resources, vegetation and ecology at the European scale. A stepwise improvement of the basic understanding of the climate system and related water fluxes can be observed. Projects like PRUDENCE and ENSEMBLES focussing on regional models and applying an ensemble of different approaches helped to quantify uncertainty and to identify possible gaps. Another set of projects, e.g. ATEAM and EEA OUTLOOK, dealt with the impacts of these changes, whereby also changes in land use, technology and demography were taken into account. Projects like HarmoniCOP and NeWater have their focus more on the social dimension of river basin planning and on adaptation to climate change. The results of the different projects have been summarised in several reports funded by the EU commission or by the European Environmental Agency.

However, the reports also reveal that many aspects of the system climate-hydrology-vegetation-society are far from being fully understood, and in some areas only little progress can be observed over the last decades. Examples in terms of climate science are the development of clouds under certain conditions, the interaction with land use pattern and changes and the development of short-term extreme precipitation events leading to high water levels and floods. Also, the role of vegetation within the climate system and the response of different types of vegetation (annual, perennial) to CO_2 fertilisation and water stress is, however investigated at the plot scale, still not adequately described in the climate and hydrological models. These processes can lead to feedbacks potentially giving a different response of the hydrological system to climate change in many regions and have therefore to be investigated further.

Moreover, the human dimension of the changes is mostly not in the focus of the investigations. Normally, human interventions and management options are considered at the level of socio-economic sectors like agriculture, water supply, power industry or fishery industry. However these sectors are built by individual actors. It might well be that a sector as such can adapt to climate change or in certain regions even gain, but the single actor, e.g. a farmer, has to make significant financial decisions and investments, which may not pay off if the climate change does not develop as expected. This is why it is not only important to have a more profound understanding of climate change at the regional scale, but also to support the adaptation process by the national and European legislation (see Section 5) to improve the adaptive capacity under uncertain climate change conditions.

A possibility and real chance to consider climate change in river basin management planes all over Europe would be the already ongoing implementation of the Water Framework Directive (WFD), where sustainable management of water resources is demanded. Reviewing the ongoing planning process though, it becomes obvious that until now climate change plays only a minor role in the planning process and needs much more attention. The EU project EURO-LIMPACS, dealing with climate change impacts on water quality, the core issue of the WFD, has the potential to bridge this gap, but further support is needed.

7.8 ACKNOWLEDGEMENTS

The work reported here was supported by the European Union (NeWater and Harmoni-CA projects), as well as the German Federal Ministry for Environment and the German Federal Environmental Agency (project "Impacts of Climate Change on Water Resources - Adaptation Strategies for Europe"). The authors would like to thank Joachim Glauer, Bärbel Uffrecht and Christian Molkenthin (PIK, Potsdam) for data support.

7.9 REFERENCES

Arnell, N.W. (1999). The effect of climate change on hydrological regimes in Europe: a continental perspective. *Global Environmental Change-Human and Policy Dimensions*, **9**(1), 5-23.

Arnell, N.W. (2003). Effects of climate change on river flows and groundwater recharge: UKCIP02 Scenarios. UK Water Industry Research Ltd. (London.

ATV-DVWK (2000). Die Elbe und ihre Nebenfluesse. Belastung und Trends, Bewertung, Perspektiven. Technical report, Deutsche Vereinigung f¨ur Wasserwirtschaft, Abwasser und Abfall e. V., Hennef.

Bronstert, A., Carrera, J., Kabat, P., and Luetkemeier, S. (2005). Coupled Models for the Hydrological Cycle -Integrating Atmosphere, Biosphere and Pedosphere. Springer, Berlin.

Cooper, D.M., Wilkinson, W.B. and Arnell, N.W. (1995). The effects of climate changes on aquifer storage and river baseflow. *Hydrological Sciences Journal-Journal Des Sciences Hydrologiques*, **40**(5), 615-631.

Döll, P. (2005). Szenarien der Wasserverfügbarkeit und -nutzung als Grundlage für eine integrative Wasserpolitik (Scenario of water availability and use as basis for integrated water policies). *Zeitschrift für angewandte Umweltforschung.15/16 (in press) scenario.*

EEA (2004). Impacts of Europe's changing climate - An indicator based assessment. EEA Report No 2/2004, EEA Technical Report no. 2/2004, European Environment Agency.

EEA (2005a). Climate change and river flooding in Europe. EEA Briefing 01/2005, EEA Briefing 01/2005.

EEA (2005b). Vulnerability and Adaptation to Climate Change in Europe. EEA Technical Report no. 7/2005, European Environment Agency.

EEA (2006). Urban sprawl in Europe - The ignored challenge. EEA Report No 10/2006, European Environment Agency.

EEA, European Environment Agency (2007). Climate change and water adaptation issues. Technical Report 2/2007.

Eisenreich, S.J. (2005). Climate Change and the European Water Dimension. European Commission- Joint Research Centre, Ispra, Italy.

Eitzinger, J., Stastna, M., Zalud, Z. and Dubrovsky, M. (2003). A simulation study of the effect of soil water balance and water stress on winter wheat production under different climate change scenarios. *Agricultural Water Management*, **61**(3), 195-217.

European Commission (2005). Communication from the Commission to the Council, the European Parliament, the European Social and Economic Committee and the Committee of the Regions: Winning the Battle against Climate Change. COM(2005) 35 final.

European Commission (2006). A European Strategy for Sustainable, Competitive and Secure Energy, GREEN PAPER, COM (2006) 105 final, Brussels, 8.3.2006.

EUROSTAT (http://eep.eurostat.ec.europa.eu/portal)

Gagnon-Lebrun, F. and Agrawala, S. (2006). Progress on adaptation to climate change in developed countries. An analysis of broad trends. OECD, Paris.

Goodess, C. (2005). Stardex - downscaling climate extremes. Statistical and Regional dynamical Downscaling of Extremes for European regions. STARDEX final report.

Hattermann, F.F., Krysanova, V., Wechsung, F., Wattenbach, M. (2004). Integrating groundwater dynamics in regional hydrological modelling. *Environmental Modelling & Software,* **19**(11), 1039-1051.

Hurrell, J.W. and Van Loon, H. (1997). Decadal variations in climate associated with the North Atlantic Oscillation. *Climatic Change*, **36**(3 - 4), 301.

IPCC (2001). Third Assessment Report on Climate Change. International Panel on Climate Change. Cambridge University Press.

IPCC (2007a). Climate Change 2007: The Physical Science Basis - Summary for Policymakers. Contribution of Working Group I to the Fourth Assessment Report of the Intergovernmental Panel on Climate Change. IPCC Secretariat.

IPCC (2007b). Climate Change 2007: Climate Change 2007: Impacts, Adaptation and Vulnerability - Summary for Policymakers. Working Group II Contribution to the Fourth Assessment Report of the Intergovernmental Panel on Climate Change. IPCC Secretariat.

Jacop et al. (2007) personal communication, see http://climate-water-adaptation-berlin2007.org/documents/jacob.pdf.

Jones, P.D., New, M., Parker, D.E., Martin, S. and Rigor, I.G. (1999). Surface air temperature and its changes over the past 150 years. *Reviews of Geophysics*, **37**(2), 173-200.

Kabat, P., Claussen, M., Dirmeyer, A., Gash, J., de Guenni, L., Meybeck, M., Pielke, R., Voeroesmarty, C., Hutjes, R., and Luettkemeier, S. (2004). Vegetation, Water, Humans and the Climate - A New Perspective of an Interactive System. IGBP. Springer, Heidelberg.

Kabat, P., Schulze, R. E., Hellmuth, M. E., and Veraart, J. A., editors (2002). Coping with Impacts of Climate Variability and Climate Change in Water Management: A Scoping Paper. Wageningen.

Klein Tank, A.M.G., Wijngaard, J.B., Können, G.P., Böhm, R., Demarée, G., Gocheva, A., Mileta, M., Pashiardis, S., Hejkrlik, L., Kern-Hansen, C., Heino, R., Bessemoulin, P., Müller-Westermeier, G., Tzanakou, M., Szalai, S., Pálsdóttir, T., Fitzgerald, D., Rubin, S., Capaldo, M., Maugeri, M., Leitass, A., Bukantis, A., Aberfeld, R., Engelen, A.F.V.v., Forland, E., Mietus, M., Coelho, F., Mares, C., Razuvaev, V., Nieplova, E., Cegnar, T., López, J.A., Dahlström, B., Moberg, A., Kirchhofer, W., Ceylan, A., Pachaliuk, O., Alexander, L.V. and Petrovic, P. (2002). Daily dataset of 20th-century surface air temperature and precipitation series for the European Climate Assessment. *International Journal of Climatology*, **22**(12), 1441-1453.

Kruger, A., Ulbrich, U. and Speth, P. (2001). Groundwater recharge in Northrhine-Westfalia predicted by a statistical model for greenhouse gas scenarios. *Physics And Chemistry Of The Earth Part B-Hydrology Oceans And Atmosphere*, **26**(11-12), 853-861.

Krysanova, V., Hattermann, F. and Habeck, A. (2005). Expected changes in water resources availability and water quality with respect to climate change in the Elbe River basin (Germany). *Nordic Hydrology*, **36**(4-5), 321-333.

Krysanova, V., Kundzewicz, Z.W., Pińskwar, I. and Habeck, A. (2006). Regional Socio-economic and Environmental Changes and their Impacts on Water Resources on Example of Odra and Elbe Basins. *Water Resources Management*, **20**(4), 607-641.

Kundzewicz, Z.W. and Schellnhuber, H.J. (2004). Floods in the IPCC TAR Perspective. *Natural Hazards*, **31**(1), 111.

Lehner, B., Döll, P., Alcamo, J., Henrichs, T. and Kaspar, F. (2005). Estimating the impact of global change on flood and drought risks in Europe: a continental, integrated analysis.

Lehner, B., Heinrichs, T., Döll, P. and Alcamo, J. (2001). EuroWasser - Model-based assessment of European water resources and hydrology in the face of global change. Kassel World Water Series 5, Center for Environmental Systems Research, University of Kassel, Kassel, Germany.

Leipprand, A., T. Dworak, M. Benzie, M. Berglund, F. Hattermann, V. Krysanova, J. Post, S. Kadner, forthcoming: Impacts of climate change on water resources – adaptation strategies for Europe. On behalf of the German Federal Environment Agency.

Luterbacher, J., Dietrich, D., Xoplaki, E., Grosjean, M. and Wanner, H. (2004). European seasonal and annual temperature variability, trends and extremes since 1500. *Science*, **303**(5663), 1499-1503.

Milly, P.C.D., Wetherald, R.T., Dunne, K.A. and Delworth, T.L. (2002). Increasing risk of great floods in a changing climate. **415**(6871), 514.

OECD (1994). OECD Core Set. Corps central de l'OCDE: Environmental Indicators. Indicateurs d'environnement. Technical report, Organisation for Economic Co-operation and Development.

Pahl-Wostl, C. (2005). The Implications of Complexity for Integrated Resources Management. In Pahl-Wostl, C., Schmidt, S., and Jakeman, T., editors, Complexity and Integrated Resources Management. International Environmental Modelling and Software Society, Osnabrueck, Germany. KeyNote Paper.

Palmer, T.N. and Raisanen, J. (2002). Quantifying the risk of extreme seasonal precipitation events in a changing climate. *Nature*, **415**(6871), 512-514.

Parry, M.L. (2000). Assessment of Potential Effects and Adaptations for Climate Change in Europe. Summary and Conclusions. T.E.A. Project., Jackson Environment Institute, University of East Anglia, Norwich, United Kingdom.

Schär, C., Vidale, P.L., Luthi, D., Frei, C., Haberli, C., Liniger, M.A. and Appenzeller, C. (2004). The role of increasing temperature variability in European summer heatwaves. *Nature*, **427**(6972), 332-336.

Schnur, R., Reichert, B.K. and Bengtsson, L. (2002). Global ocean warming tied to anthropogenic forcing. *Geophysical Research Letters*, **29**(11).

Schröter, D., Acosta-Michlik, L., Arnell, A.W., Araújo, M.B., Badeck, F., Bakker, M., Bondeau, A., Bugmann, H., Carter, T., de la Vega-Leinert, A.C., Erhard, M., Espiñeira, G.Z., Ewert, F., Fritsch, U., Friedlingstein, P., Glendining, M., Gracia, C.A., Hickler, T., House, J., Hulme, M., Kankaanpää, S., Klein, R.J.T., Krukenberg, B., Lavorel, S., Leemans, R., Lindner, M., Liski, J., Metzger, M.J., Meyer, J., Mitchell, T., Mohren, F., Morales, P., Moreno, J.M., Reginster, I., Reidsma, P., Rounsevell, M., Pla, E., Pluimers, J., Prentice, I.C., Pussinen, A., Sánchez, A., Sabaté, S., Sitch, S., Smith, B., Smith, J., Smith, P., Sykes, M.T., Thonicke, K., Thuiller, W., Tuck, G., van der Werf, G., Vayreda, J., Wattenbach, M., Wilson, D.W., Woodward, F.I., Zaehle, S., Zierl, B., Zudin, S., Cramer, W. (2004) Advanced Terrestrial Ecosystem Analysis and Modelling (ATEAM) Final report 2004. www.pik-potsdam.de/ateam/.

Schröter, D., Cramer, W., Leemans, R., Prentice, I.C., Araujo, M.B., Arnell, N.W., Bondeau, A., Bugmann, H., Carter, T.R., Gracia, C.A., De La Vega-Leinert, A.C., Erhard, M., Ewert, F., Glendining, M., House, J.I., Kankaanpää, S., Klein, R.J.T., Lavorel, S., Lindner, M., Metzger, M.J., Meyer, J., Mitchell, T.D., Reginster, I., Rounsevell, M., Sabate, S., Sitch, S., Smith, B., Smith, J., Smith, P., Sykes, M.T., Thonicke, K., Thuiller, W., Tuck, G., Zaehle, S. and Zierl, B. (2005). Ecology: Ecosystem service supply and vulnerability to global change in Europe. *Science*, **310**(5752), 1333-1337.

Soncini-Sessa, R. (2004). Integrated modeling and participatory decision making in practice: The Verbano project. Mc Graw Hill, Milan. In Italian, English forthcoming with Elsevier.

TAC (2000). Integrated Water Resources Management. TAC Background Papers no. 4, Global Water Partnership Technical Advisory Committee.

UNCSD (1996). Commission on Sustainable Development: Indicators of Sustainable Development Framework and Methodologies. Technical report, United Nations.

van Beek, E., Hellmuth, M., Schulze, R., and Stakhiv, E. (2002). Coping with Climate Variability and Climate Change in Water Resources. In Kabat, P., Schulze, R., Hellmuth, M., and Veraart, J., editors, Coping with Impacts of Climate Variability and Climate Change in Water Management: A Scoping Paper. International Secretariat of the Dialogue on Water and Climate, Wageningen

Wilby, R.L., Orr, H.G., Hedger, M., Forrow, D., and Blackmore, M. (2006). Risks posed by climate change to the delivery of Water Framework Directive objectives in the UK. Environment International, **32**(8), 1043-1055.

8

Monitoring information systems to support integrated decision-making

Raffaele Giordano, Vito F. Uricchio, Michele Vurro

8.1 INTRODUCTION

A decision-making process is widely considered as an iterative process aimed at identifying the most suitable alternative to solve a given problem. Several models have been developed to describe the decision process. Most of them assume that the decision process consists of different steps, such as the PrOACT paradigm (Raiffa *et al.* 2002). According to this model a decision process goes through the steps of identification of the problem and objectives, generation of alternatives, evaluation of consequences and making tradeoffs. Generally speaking, decision-making is largely considered as a rational process based on the use of all available information. Therefore, the information can support different phases of the decision process, ranging from the problem definition, to the alternatives construction and evaluation. The information plays a fundamental role to improve the understanding of the consequences of, and trade-off among, the alternatives (McDaniels and Gregory 2004). The information reduces the uncertainty and facilitates the definition of alternative scenarios, supporting the decision makers to define the possible outcomes of alternatives, the likelihoods and the resulting consequences of these outcomes (Raiffa *et al.* 2002).

These broad concepts should also be implemented in decision processes dealing with environmental resources management. Policies and decision-making processes on environmental issues need to be based on appropriate information and modelling results, as is widely acknowledged in literature. Sound knowledge, reliable information and accurate data are vital for good environmental decision-making (Haklay 2003).

Increasing knowledge of the complexity of the processes in water systems has led to a growing demand for environmental information (Timmerman *et al.* 2000). Over the last two decades, water management has changed from a focus on distinct sectors, to a more integrated perspective. To construct and select the most suitable alternatives for water management, a holistic approach needs to be adopted. In the water resources management domain, there is increasing interest in integrated assessment that is emerging as a new discipline that integrates different kinds of knowledge to be used in all phases of the decision process. Water management is a complex process involving many actors, with different knowledge and at different levels. The success of management initiatives in such a context depends on their ability to facilitate co-operation between the actors and integration between different sources of knowledge. Thus, the flow of information between the actors, and an analysis of the ways in which networks of actors and institutions communicate internally and externally is vital (Timmerman and Langaas 2004).

This chapter discusses how monitoring and information can support decision-making in the current complexity of IWRM. It reviews the results of EU projects dealing with this subject. The review aims to verify if and how the research in the monitoring field is fulfils the IWRM requirements of information production and management.

The chapter is organised as follows: This section describes the role of information to support learning in IWRM and the challenges to information production and management coming from the increasing awareness of complexity in water management. In Section 8.2, the results of EU funded projects review are described. Finally, in Section 8.3, conclusions are drawn from this review together with suggestions for further research in the environmental monitoring field.

8.1.1 Complexity and IWRM – The role of learning

Traditional approaches to decision analysis focus mainly on the generation and evaluation of alternatives, given a present state and a desirable state. The alternatives are considered as means to reduce the differences between these two states, to transform the current system into the target configuration (Rosenhead and Mingers 2001). Traditional approaches consist of systematically ordered thinking concerned with means-definition in well-structured problems in which desirable ends can be stated (Rosenhead and Mingers 2001). They assume the decision makers are interested in using the best available information to be used for quantitative and qualitative evaluation and for the comparison of the alternatives (McDaniels and Gregory 2004). The traditional approach assumes that decision-making is largely a rational process in which the individuals order the possible outcomes of a decision situation and choose the strategy according to their preferences (Timmerman and Langaas 2004) and to the available information.

Conversely, in a real situation, making decisions takes place in conditions of complexity, in which the most demanding and troublesome task is to define the nature of the problem, rather than its solutions (Rosenhead and Mingers 2001). Managing means interpreting and reacting to the flux of interacting events and ideas of the real world (Checkland 2001). In this perspective, there is no unique definition of the problem, but each individual has his own perspective in defining and interpreting a problem situation (Lane and Oliva 1998). During the decision process, individuals continually negotiate and re-negotiate with others their perceptions and interpretations of the real world outside themselves, leading to a social construction of the problem domain (Pahl-Wostl 2007). The expected outcome is a set of insights that emerge from the comparison of individual perspectives (Lane and Oliva 1998), forming the richest possible picture of the problem situation (Checkland 1981).

A distinction is made between hard and soft system thinking, whereas the former adopt an "objectivist" stance that sees problems as independent of individual's views and beliefs. Soft system thinking requires a "subjectivist" stance that recognises the importance of participants' perceptions (Rosenhead and Mingers 2001), and thus of their mental models. Mental models influence an actor's perception of a problematic situation by influencing both his/her observation of the world and his/her conclusions based on observations (Pahl-Wostl 2007). They can be considered as the window through which people view the world (Timmerman and Langaas 2004). Mental models determine what data the actors perceive in the real world and what knowledge the actors derive from it (Kolkman *et al.* 2005).

Thus, the use of information in decision-making is rational only to a certain point. It will not be possible for any decision-makers to deal with a problem situation adopting a neutral, objective approach. His/her personality traits, experiences, knowledge and interests will affect what is noticed and what is taken to be significant (Checkland 2001). Differences in mental models cause communication barriers that prevent mutual learning and understanding (Kolkman *et al.* 2005). This lies at the basis of conflicts between different actors involved in a decision process. Thus, in a complex world the use of information is limited by mental models, and therefore people that do not understand decision-makers' mental models may not understand the rationale of decisions. Only when information fits mental models, can it be used to support decision-making processes (Timmerman and Langaas 2004). Thus, traditional models, if they are to describe decision processes, need to be extended taking into account the role of individual mental models. It does matter in information production and management whether, or not, there is a clear understanding and sharing of information users' mental models.

Different methodologies have been developed to define decision makers' mental models (Checkland 2001; Timmerman forthcoming; Kolkman *et al.* 2005). Most of them imply the involvement of a wide variety of information users (i.e. scientists, managers, policy makers and public at large) in a debate in which assumptions about the world are teased out, challenged, tested and discussed (Checkland 2001). During the debate, participants become aware about each other's perspectives and key interests. The objective of the debate is the establishment of a common understanding, which supports information exchange and co-operation.

The existence of different mental models influences which information is collected and accepted. Once information has been examined, a perspective is developed, and an insight is gained and integrated into the mental model itself (Kolkman *et al.* 2005). Information may prove initial mental models to be wrong and support the debate between actors, which may lead to a revision of mental models, through reflection and negotiation, in a social learning process. This learning may support changes in the water management mental model. Moreover, feedback on management actions may generate new questions or new insights, making the originally agreed upon information appear inadequate. Changes in the environment or society, possibly influenced by management measures, may stimulate novel insights that raise new information needs that can lead, ultimately, to new policies. In addition, the information users' perspectives may change, influenced by the evolving knowledge of the participants of the issues at stake through the learning process, resulting in new information needs. Changes in water management mental model influences what new information the actors care to observe in reality. Thus, the information needed to support a decision process evolves according to the actors' learning process, leading to revision/adaptation in monitoring strategies and data interpretation.

Recognition of the importance of mental models and learning has led to the introduction of the Adaptive Management (AM) concept. AM refers to a "learning by doing" process in which the outcomes of the implemented strategies are used to refine iteratively, and to improve, management policies. Adaptive management proceeds from the premise that policies can be treated as experiments in which monitoring and evaluating outcomes, and judging what has been learned are fundamental steps (McDaniels and Gregory 2004). Rather than using existing knowledge to implement a single best plan for ecological management, AM approaches explicitly recognise the existence of uncertainty and propose that a range of management approaches be tested and refined over time based on a careful comparison of results.

The implemented actions become part of the world leading to new perceptions of the problem situation, to new evaluation and further actions (Checkland 2001). Information collected during the implementation increases understanding of the system and supports a revision of the initial hypothesis based on the actors' mental models. Emphasising the role of learning in AM leads to a focus on the role of feedback information from the implemented actions, as described in the double-loop learning model developed by Sterman (2000). In one loop a decision is made, and the results are used to support further refinement. Feedback on actions can also induce changes in the understanding of the system (Fazey *et al.* 2005). Such a feedback-based learning model stresses the need for monitoring the discrepancies between intended and actual outcomes (Fazey *et al.* 2005).

The increasing awareness of the complexity of environmental systems and of management as learning rather than control (Pahl-Wostl, 2007) is changing the role of information in decision-making. To support the learning process, information concerning both negative and positive feedback of management actions are required for the iterative evaluation of both the continued desirability of objectives, and progress toward their achievement (Lessard 1998). Corresponding adjustments are made to implementation activities and management objectives.

Information should support actors in identifying possible changes in the system and to evaluating whether they occur because of the implemented actions. Information has to be "action-oriented", concerning the *effectiveness* (i.e. if the implemented action is "the right thing to do"), the *efficacy* (i.e. if the actions actually work towards the goal or not), the *efficiency* (i.e. informing the actors about the sustainability of actions with regard to resources used) (Checkland 2001)

In this perspective, information becomes the primary tool for learning about the system and assessing the management strategies. In the next section, the IWRM and AM requirements regarding information production and management are described.

8.1.2 Complexity and information production and management

Given the role of information in supporting decision making in a complex world, novel monitoring information systems are needed. To support the learning process, a monitoring system cannot be considered as a simple bin of information, but it should be seen as part of the learning process itself. However, if it is to play this fundamental role, a monitoring system needs to overcome the shortcomings of the current practices (Giordano *et al.* forthcoming).

Monitoring is often understood in water management as a rather technical task. The design of monitoring networks is, as a consequence, based on general monitoring objectives that ignore what information could be relevant for a management process. Without clear monitoring design information objectives, the focus was mainly on aspects of data collection, i.e. the choice of sensors, the distribution of the monitoring stations, the distribution over time of the data, etc. This type of technical perspective, developing large, high-tech and expensive monitoring networks, has historically dominated monitoring designs (Giordano *et al.* forthcoming). In this perspective, i.e. the more the better, or money-in-data-out, the efficacy of monitoring networks was mainly evaluated using easily measurable but often irrelevant standards: the quantity of stations, or the use of highly sophisticated equipment, etc.

To support learning in water management, monitoring must serve an information purpose to support action (Enderlein 2004). A change in focus is needed from data to information and from knowing what and how, to knowing why. The quality of information increases as it meets policy-related needs. Therefore a major task in designing a monitoring network is to evaluate the management's needs to determine the type of information the monitoring system is to acquire. This includes a discussion among information users and monitoring system designers regarding not only what the user wants, but also what the monitoring system is capable of producing (Ward 1986). Improved co-operation between information producers and users makes overcoming the so-called "water information gap" easier (Timmerman and Langaas 2004) and increases the comprehensibility of information. Moreover, the level of acceptance of the information generated improves (Denisov and Christoffersen 2001).

This cannot be done without the establishment of an iterative dialogue, well represented in the information cycle, between information users and producers (Timmerman *et al.* 2000). In this framework, information needs are the link between information and the decision process. Specification of information needs is a means to make a translation from a policy problem into an information management issue; water policy and management objectives are translated into information expectations that, in turn, form the basis for an information production network.

Currently, monitoring system outcomes are often only understandable by experts and therefore not "accessible" for a wider audience, limiting public participation in decision-making processes. These shortcomings of traditional monitoring systems are also described in the usability problem literature (e.g. Keenan *et al.* 1999). Problems are mainly related to human / information system interactions. One of these problems is the so called "water information gap" between the sources and the receivers of information (Timmerman and Langaas 2004). Information producers often do not speak the same language as information users. The information gap is even deeper when the information users are participants in a public participation process. The participatory aspect of water management, as required by the EU Water Framework Directive, requires appropriate access to information concerning water resources for all the parties interested in the decision-making. The quality of public participation in environmental decision-making improves as the public gains access to environmental information, data and knowledge (Haklay 2003; Gouveia *et al.* 2004).

Public access to environmental information has also been emphasised by the Aarhus Convention. The Aarhus Convention establishes a number of rights of the public with regard to the environment. Concerning the information, the convention establishes the right for everyone to have access to environmental information held by public authorities. This can include information on the state of the environment, but also on policies or measures taken, or on the state of human health and safety where this can be affected by the state of the environment. The convention also establishes the right to participate in environmental decision-making.

Thus, an important difference between traditional monitoring systems and those able to support learning regards the information users. In fact, while traditional monitoring systems collect information mainly for water managers, following a top-down approach to environmental management, novel monitoring systems should be based on the direct involvement of a wide variety of information users, i.e. scientists, water managers, policy makers and the public at large (Timmerman forthcoming). Therefore, a monitoring information system must be designed to facilitate data access, visualisation and communication. The information provided has to be easily comprehensible even for non-expert people. That is to say, it is fundamental to develop different interfaces with different level of access according to different user profiles. The integration of a monitoring system with Information and Communication technologies has great potentialities in improving environmental information access.

In a complex system, decision-makers are required to operate in an environment of densely interconnected networks, in which the wider ramification of decisions cannot be ignored (Rosenhead and Mingers 2001). Facilitating the flow of information between the actors involved in decision processes through these networks is crucial for the success of management initiatives (Timmerman and Langaas 2004). Monitoring in the past tended to be divided between, and even named after, different disciplines. Too many people involved in water monitoring seem to be blinded by their disciplines, seeing the monitoring system as an extension of their science and technology (Ward 1994). To deal with complexity in water management requires information production and management systems based on gathering, synthesising, interpreting and communicating information from various domains and disciplines. There is a need to make a shift from multi-disciplinarity to inter-disciplinarity, based on close and regular contact between disciplines (Timmerman and Langaas 2004).

A novel monitoring information system should be based on the dialogue between all the actors involved in information production, management and use. The aim of this dialogue is to reach a common understanding about the information needed to support decision-making.

During this debate, the participants' mental models can be developed. Two mental models can be defined, i.e. "water management mental models" and "information management mental models", where the former concerns the interpretation of the problem at a certain stage and the latter, the information needed to solve the considered problem, and the "frames" used to interpret the information (Pahl-Wostl 2007; Kolkman *et al.* 2005). These mental models should be adopted as a basis for the monitoring system design. The variables included in these models are the information needed to monitor the effects of water management actions on the system. The links between the variables allow the monitoring system to be designed taking into account the strong interrelationships between the different parts of a complex system (Giordano *et al.* forthcoming).

Two learning processes can be supported by a monitoring system. The first concerns the water management, as described in the previous section. The second learning process relies on feedback from applied monitoring practices. As a result of experience in implementing the monitoring program and assessing its results, adaptation to monitoring may be needed (Cofino 1994; Smit 2003). Therefore, the development and revision of participants' mental models should be used as basis for monitoring design and revision (Giordano *et al.* forthcoming).

8.1.3 Complexity and monitoring system

Monitoring and decision-making processes cannot be considered in parallel, but there is a need for a strong and continuous interrelationship. The monitoring program needs to be adapted to the different stages of the policy definition process because each stage requires other types of information (Cofino 1994). Where current monitoring is adapted after changes in policy or management are implemented, adaptive water monitoring should be an integral part of the water management process as well as the water governance process.

The concept of an Adaptive Monitoring Information System (AMIS) can now be introduced. An AMIS supports the learning processes both in water management and in monitoring strategies. An AMIS adopts an adaptive approach to monitoring design, which iteratively refines the monitoring design as a result of experience in implementing the monitoring program, assessing its results and interacting with the users. This observation actually describes a learning process in monitoring practices. This learning process requires adaptation of the information management mental model. Revision/adaptation can occur in information needs, monitoring strategies and data interpretation due to changes in mental models.

The difference between the current approach to monitoring design and an adaptive approach is that, in the adaptive approach, the barriers to monitoring design are overcome by adaptively implementing monitoring, rather than waiting for new information or designing a system that does not anticipate new information (Ringold *et al.* 1999). The development of a monitoring network is, in this situation, a learning process that can, for instance, enhance policy development and management effectiveness (Giordano and Vurro 2005). In this learning process, negative feedback shows the need for adaptation of the monitoring programme. Partly, the causes can be found within monitoring practices:

- Too little attention may have been given to specifying the information needs;
- The information needs may have been specified in such a way that no adequate information could be produced from it or so that it does not reflect the actual information users' needs;
- The selected indicators may not adequately measure what they are purported to measure, or the strategy to collect information may not have produced the right information.
- The available budgets may restrict the number of indicators that can be measured or the intensity of the network in terms of locations and frequency.

The causes for negative feedback can also be found in the management actions:

- The information obtained may generate new questions or new insights making the originally agreed upon information appear inadequate;
- Environmental conditions, possibly influenced by management measures, and/or societal conditions may change;
- As a result, combined with changing insight, policies may change, leading to changing information needs;
- The information users' perspectives may change, influenced by the evolving knowledge of the participants of the issues at stake through the learning process, resulting in new information needs.
- New methods may have been developed that provide information which until recently was not available (e.g. Giordano and Vurro 2005; Timmerman *et al.* 2000).

Therefore, the properties of an AMIS have to be defined taking into account the role of the learning process both in water management and in monitoring. In Table 1, a comparison between the conventional approach and monitoring to support IWRM and AM is proposed.

Table 1: Comparison among current, IWRM and AM monitoring

Current monitoring practices	Needs for IWRM
Based on monitoring objectives and disciplinary needs Information users have unrealistic expectations of the information that will be produced Data accessibility is limited Abundant and detailed information is provided The information provided is highly specialised The available information is divided over various organisations Information is transferred to the information users	Based on policy objectives and information users' needs The information that will be produced is jointly agreed between information users and producers Data are publicly available and accessible The information provided is concise and addresses the policy objectives The information is targeted towards specific audiences The information combines results from various organisations and is integrated over disciplines Information is communicated to the information users and a broader stakeholder or public audience and evaluated before being incorporated into policy support
	Additional needs for AM
The outcomes of the monitoring program (data) are the focus. The purpose of the monitoring program is to evaluate environmental status set against target values. Monitoring follows management and policy implementation.	The monitoring program design and the responses on this design are as important as the results: the focus is on learning. Monitoring becomes the primary tool for learning, i.e. understanding the system, assessing the effectiveness of management activities evaluating the system changes, and measuring progress towards participatory defined goals. Monitoring, management and governance are interdependent.

8.1.4 Review framework

This chapter is based on a review of several research projects dealing with monitoring information systems to support IWRM. The review process was carried out by scoring the different projects according to specific IWRM criteria. The criteria were identified considering the properties of monitoring systems able to deal with complexity in water management, as described in the previous sections.

As previously discussed, to deal with complexity monitoring systems have to be linked with the information management process, and together they need to be interconnected with the decision process. Thus, the capabilities of monitoring systems to produce useful information to support a decision process were taken into account in carrying out the review of EU funded projects. Moreover, the ability of monitoring systems to enhance information accessibility in order to support public participation was considered.

Another important characteristic is the adoption of an inter-disciplinarity approach to information collection and management. Furthermore, due to the ramification of decisions in a complex world, monitoring information systems should support the information exchange between all the involved actors.

During the review process, four main themes were considered to describe the results of the EU funded projects considered, namely:

- Interconnection between monitoring/information management and decision process;
- Information exchange among actors;
- Integration of different types of information;
- Degree of information accessibility for both the decision-makers and stakeholders.

The results of the review are described in the following section.

8.2 REVIEW OF EU FUNDED RESEARCH PROJECTS

The review process aims to verify whether these themes have been included within the objectives of each project. The selection of the projects was carried out consulting the Cordis database. All the projects focused on integrated water management and monitoring were considered. The review process regarded both projects focused on monitoring issues and projects that address monitoring issues as part of another topic. The technical issues related to environmental monitoring are not considered in this review, therefore, projects dealing only with these issues were excluded.

Nine projects were selected, as described below:

- **Mantra – East Project** (http://mantraeast.webmedia.ee/)**:** The Mantra – East project aimed at analysing and developing strategic planning methodologies and scientific tools to support IWRM in transboundary river basins. One of the four modules of the project addressed the issues of environmental information for Policy and Decision-makers and one of scientific objectives regards the improvement of information utilisation in environmental resources management. Thus, the importance of an information management/decision process link has been recognised and explicitly assumed as an objective of the project. Furthermore, Lake Peipsi, one of the case studies, has been recognised as a combination of "data rich" and "data poor" with respect to environmental information, and the access and availability of data is scattered. A monitoring system was designed that considers eutrophication as one of the most pressing environmental problem.

- **Tisza River Basin Project** (http://www.tiszariver.com/)**:** The general objective of the project was to save water resources and ecological values with the help of integrated catchment management tools and to secure the sustainable use of the resources of the Tisza River Basin. The scientific objective was to develop a "real-life-scale" integrated catchment model system. In this the term "real-life scale" means the development (and selection) of a practical application-oriented set of tools, (i.e. a set of computer models for water flow, water quality and ecosystem functioning), that are exactly tailored to the issues to be solved and the availability of data. The establishment of an international database of the Tisza River Basin was also a major objective. The project proposal lays special emphasis on an ecohydrological approach to the revitalisation of the broad, numerous and unique wetlands and oxbow lakes, along the Tisza River and its tributaries.
- **TransCat Project** (http://www.transcat-project.net/)**:** The main goal of the project is to create an operational and integrated comprehensive Decision Support System (DSS) for optimal water management of catchments in borderland regions. More specifically the project's objective is the design and development of a prototype for an operational integrated DSS of a comprehensive character, including multilingual facilities, interactive visualisation interface, the possibility of plugging in numerical models, particularly of the cross-border water resource system. The system will be tested for the conditions of the transboundary pilot areas. Another objective of the project will be the establishment of a European platform for initiating and promoting international co-operation and networking that will allow a more detailed insight to transboundary water/environment and related socio/economic problems.
- **Harmoni-CA Project** (http://www.harmoni-ca.info/)**:** The overall objective of the large scale concerted action 'HarmoniCA' is to create a forum for unambiguous communication, information exchange and harmonisation of the use and development of ICT-tools relevant to integrated river basin management, and the implementation of the WFD. HarmoniCA is a concerted action, meaning the project's innovation is based on reviewing, elaborating and synthesising available knowledge, resulting in new insights, achieving a common understanding of problems, methods and solutions. Though much of the work involves ICT-tools, HarmoniCA is not about ICT-tools as such: HarmoniCA is about harmonisation and guidance on proper development and use of ICT-tools in the light of effective and efficient development of integrated river basin management plans and the implementation of the WFD.
- **AquaStress Project** (http://www.aquastress.net/)**:** This project develops stakeholder driven, European scale, comprehensive multisectorial, integrated (institutional, socio-economic, technical) approaches for the diagnosis and mitigation of water stress. AquaStress is aimed at empowering actors at different levels of involvement, at different stages of the planning process, and working at different spatial scales, to mitigate water stress. AquaStress is aimed at leading to prioritise actions that will allow a gradual improvement and flexibility to adapt to change in global systems, knowledge, technology and society. AquaStress will deliver guidelines to implement integrated water stress mitigation options at a local, regional and European scale. AquaStress will deliver a "culture change" in approaches to water stress.
- **Combining systematic and participatory approaches for developing and promoting strategies for sustainable land and water management Project** (http://cordis.europa.eu)**:** this project deals with the integration of local and scientific knowledge for environmental management in semi-arid regions, using a GIS-based information system as the integration platform.
- **Harmoni-COP Project** (http://www.harmonicop.info/)**:** The aim of the HarmoniCOP project is to increase the understanding of participatory river basin management planning (RBMP) in Europe. RBMP is the integrated cross-sectorial planning and management of river basins if necessary across political and administrative borders. The project's objective is to generate practical information about participation processes in river basin management and to support the implementation of the public participation provisions of the European Water Framework Directive.
- **Harmoni-RIB Project** (http://www.harmonirib.com/)**:** The overall goal of HarmoniRiB is to develop methodologies for quantifying uncertainty and its propagation from the raw data to concise management information. Thus, the HarmoniRiB project is aimed at supporting the WFD implementation, by addressing issues of uncertainty in data and modelling, and by developing a 'virtual laboratory for modelling studies'. This virtual laboratory will comprise of a set of river basins, of which data relevant to modelling and the WFD

implementation are readily available for the scientific community. The data can be used for comparison and demonstration of methodologies and models relevant to the WFD.

- **NeWater Project** (http://www.newater.info/): The overall goal of NeWater project is to deal with challenges in water management due to the increasingly complexity. The project recognises the value of highly integrated solutions and advocates integrated water resource management (IWRM) concepts. However, NeWater is based on the hypothesis that IWRM cannot be realised unless current water management regimes undergo a transition towards more adaptive water management. NeWater identifies key elements of current water management regimes and investigates their interdependence. Research is focused on transformation processes of these elements in the transition to adaptive integrated water resources management.

All these projects were reviewed according to the criteria described previously. The results of the review process are shown in Table 2 and described in detail in the following pages.

Table 2: Results of the review process

Project	Monitoring and decision process	Information exchange	Integration of different types of information	Accessibility and comprehensibility
Mantra - East	++	++	++	++
Tisza River Basin	++	++	++	+
TransCat	+	++	+	+
Harmoni-CA	+	-	++	+
AquaStress	++	+	++	++
Combining systematic...	-	-	++	-
Harmoni-COP	-	+	+	++
Harmoni-RIB	-	-	++	+
NeWater	++	++	++	++

++ Explicitly considered; + Implicitly considered; - Not considered

8.2.1 Interconnection between monitoring/information management and decision process

As discussed previously, one of the basic requirements in designing an effective monitoring system to support IWRM is a close connection between the monitoring process and the decision-making process. The information produced through monitoring has to support actions in environmental management. In this section, EU funded projects aimed at designing a monitoring system based on information needs elicitation are reviewed (Table 2).

In the **Mantra-East** project, the establishment of a link between information production and management, and decision processes is one of the main objectives. The focus is on environmental information management conceived as a dedicated information- related activities aimed to guide and support policy and decision making in water management. As told previously, this project deals mainly with transboundary water management issues. Environmental information are considered as the "oil in the transboundary water management machinery" (Roll and Timmerman 2006). That means that consistent and relevant information on river basin is fundamental to support cooperation between countries. Timmerman and Langaas (2004) stated that different actors can have different perceptions about which information can be considered consistent and relevant according to their mind-frame. Thus, in the Mantra-East project the information needs of Estonian-Russian Transboundary Water Commission have been studied. The goal of the study was to characteristics of information used, i.e. type of information, presentation of information, accuracy and reliability, information producers, level of aggregation. From these experiences, it was concluded that the involvement of stakeholders in information needs definition support the learning process concerning the interrelationships between the different factors influencing the transboundary river basin management, particularly between driving forces and pressures. Moreover, the debate between participants and experts facilitate the exchange of knowledge about how to find solutions to water problems. Using the results of information needs definition process, recommendations to develop a more effective information and communication strategies for Lake Peipsi were made and partially implemented.

The **Tisza River Basin** project is aimed at establishing a production and exchange of data and information really oriented towards management issues (Jolankai 2004). To define these issues the needs of all the stakeholders in the river basin have been elicited through end-users workshops. The stakeholders involved came from the different countries and were mainly at national levels. No local levels stakeholders have been involved. In this project, the information needed by decision makers has been not used to define and implement an information strategy. The information needs have been used to identify suitable models to be used in the Tisza river basin. In fact, one of the main objectives of the project is the adoption of a "real-life-scale" approach, which means that all tools to be developed must be robust, simple and practical, with due concern to the current availability of data. Thus, in the project's perspective, monitoring must provide data both to calibrate the models and as input for them. To reach this aim, the needs of the stakeholders have been used to identify the water and environmental issues of the national involved. According to them, an international database has been developed for all type of GIS, monitoring and other data. Next step was to develop, select and adapt a set of modelling tools suited to environmental issues and data availability. From this project it can be concluded that complex models developed only within the scientific community require knowledge and data which are often not available, making these models unsuitable to support decision making process. Thus, the definition of these tools has to be based on an in depth analysis of decision makers requirements and knowledge, and on data availability. This is particularly true in areas in which, due to scarce economical and human resources, the monitoring system cannot be easily enhanced.

The main goal of the **TransCat** project is the creation of technical tools to support Integrated Water Management in transboundary river basins, in the context of implementation of the EU Water Framework Directive. The information from monitoring is used as input for the DSS. Therefore, linking monitoring information to the decision process has been considered as an important objective. The characteristics of the DSS have been established considering the water management issues in the different Case Studies, the legal requirements and stakeholders needs and goals concerning transboundary water management and collaboration development. Concerning last point, the policy objectives of the stakeholders in the different Case Studies have been defined analysing their role in the regions. Cooperation with stakeholders supports the development of DSS prototype. Once the DSS has been defined, the data requirements for DSS implementation have been identified. The analysis of DSS requirements was divided in three main categories: data collection requirements, data methods requirements, end-user requirements (end-users of the DSS should be able to evaluate the quality of input data as basis of the DSS). The analysis has been carried out considering the specificities of the different Case Studies. Therefore, also from this project it can be concluded that tools to support decision making process in complex domain, like water management, has to be shaped to the requirements of the users. Moreover, the strict interaction between experts and users during the tool development process facilitate both the process of learning and of knowledge structuring.

One of the aims of **Harmoni-CA** is the promotion of the joint use of modelling and monitoring, bridging the gap between the modelling community and the monitoring community (Kamphorst *et al.*2005). The importance of data and information need elicitation is recognised, but it is mainly related to the requirements of the models. Harmoni-CA concludes that it is fundamental to overcome the distinction between monitoring for state assessment and monitoring for modelling, responding to an integrated water management approach. There is a need for interdependency between monitoring and modelling. In fact, models suggest which indicators may be critical to monitor by posing elaborate hypotheses of what variables and relationships are critical to understanding the problem in question and then to consider the dynamic implications of these hypotheses through the simulation of different scenarios. This increases the potential to usefully extract more information from the available data and integrate monitoring data, enhancing information utilisation.

The main objective of the **AquaStress** project is to identify mitigation measures aimed at reducing water stress in several case studies. Therefore the monitoring and evaluation of implemented strategies can play an important role. Monitoring is not considered as an initial support to identify the possible actions, but is strictly linked to the decision process. Different kinds of monitoring systems will be studied and developed in this project to assess the efficacy of mitigation strategies and to suggest possible changes. There are no definitive results. Nevertheless a preliminary experimentation to define monitoring system involving the different stakeholders has been carried out in one of the Case Studies. During a workshop different cognitive maps have been defined interacting with stakeholders groups, in order to identify the variables that are needed to monitor the efficacy of water stress mitigation options. One of the most interesting conclusions from this experience concern the usability of group modelling techniques to support the identification of stakeholders information needs. In fact, the cognitive maps developed during the workshop have been used to identify the most important actions according to the stakeholders' opinion. Moreover, the cause-effect links of the aggregate cognitive map will be used to recognise the effects to be monitored and to define the monitoring program.

A similar approach is adopted in **NeWater** project. One of the objectives of this project is the development of a monitoring system able to support Adaptive Water Management, also in data poor regions. To this aim, a cognitive modelling approach has been adopted, involving stakeholders in two different Case Studies. The objective is to define the main water management problems in the regions, and the information needed to support the decision process. In one of the case study, a causal model concerning the soil salinisation has been developed considering the local knowledge. In this model, concepts from scientific literature and from the expertise of local communities have been integrated. The aim is to define a monitoring system able to collect and validate the local knowledge concerning the soil salinisation process.

The integration between monitoring and the decision making process is considered as fundamental for many EU funded projects, even if this topic has been addressed in different ways. Some projects adopt a direct approach decision-making – information production and management, while others introduce an interface, i.e. models and DSS. In the first case, the adopted approaches are based on the direct involvement of stakeholders in the definition of information needs that can be used as basis for monitoring information systems design and development. In the second case, the interface properties have been defined considering the requirements coming from decision-making field. Consequently, the data and information required have been defined.

8.2.2 Information exchange among actors

IWRM is based on the assumption that water management is a complex process, with many actors at different levels. Complex networks of individuals and organisations can be distinguished in these situations. Co-operative and conflicting mechanisms can characterise the relations between actors. The success of management strategies depends on communication of information during the management process. The flow of information plays a more crucial role in transboundary river basin management.

One of the most interesting EU funded projects on this topic is the **MANTRA – East** project. The enhancement of communication and information exchange, within and across the border, in transboundary water management is the main objective of this project. According to the conclusion of the project, transboundary water basins need to be managed interactively, i.e. through regular communication and consultation among all the major stakeholder groups in all the countries involved. Stakeholders' involvement results in recommendations to facilitate the information exchange in transboundary water management (Roll and Timmerman 2006): information should be presented in the form of maps; information should be reliable and accurate according to decision-makers perceptions; cause-effect relationships should be presented. Overall, facilitating information exchange in transboundary management implies the collection of decision makers' viewpoints for all the involved countries.

Enhancement of cross-boarder information exchange is one of the main aims of the **Tisza River Basin** Project. Some aspects of the project are interesting for the aim of this chapter. Firstly, a unique international database was developed in order to enhance the exchange of information among the different countries sharing the river basin. In this database all the data coming from different countries are stored. This common database includes GIS based maps (such as digital terrain models, land use maps), time series of monitoring stations and other data, also relaying on space imagery and land reference measurement techniques. A web-based GIS system (Tisza River Information System - TRIS) has been developed considering data coming from the different countries. The methodological approach for the establishment of TRIS comprised: examination of existing data bases with respect to homogeneity, correctness, and plausibility; the setup of a GeoData server for storage, processing and analysis of spatial data together with relational data; implementation of interfaces enabling the online access via Internet; input of vector and raster GIS layers, time series and descriptive data into the project's database. The project concludes that the information exchange between actors in transboundary management can be improved developing an information system in which the knowledge and data coming from different counties can be integrated according to shared protocols. Anyway, this project demonstrates that this tool can be successful only if actors from all the countries in the basin are involved in the development process from the beginning.

From an information exchange point of view, the **TransCat** project focuses on the definition of standards for data collection and management (Bender *et al.* 2004). In fact, the basis of the DSS for transboundary water management is a database, containing all available data and information in border regions. The development of such a database implies the integration of data and information coming from national monitoring networks. Often, these data may differ in accuracy, precision, sampling intervals, completeness, detection limits and scale. To facilitate the integration among these data, the norms and regulations for each country have been collected and evaluated allowing a comparison of different methods of data collection and management. The evaluation of these methods has led to the creation of a manual for standard procedures of sampling and analysis to be used in the countries involved. The creation of these protocols has been considered fundamental to facilitate the data input and integration in the DSS.

NeWater aims to facilitate the exchange of information among actors involved in water managements. This issue has been addressed both analysing the flows of information between countries sharing transboundary river basins, and developing monitoring and information management systems according to information needs of all decision makers involved in the decision process. Concerning the first point, a preliminary conclusion that can be drawn from the experience regards the difficulties in supporting information exchange between countries with low or even no diplomatic relationships. In these situations, even if some transboundary institutions exist, their importance in influencing the water management strategies is limited. Concerning the development of a monitoring and information management system, a stakeholders based approach has been adopted. The main goal is to create a common understanding about the information required for the decision making process. Moreover, a shared GIS will be developed cooperating with the information users to facilitate the integration between information coming from different information providers.

The exchange of information between actors is implicitly considered as an objective for AquaStress and Harmoni-COP. In **AquaStress** the definition of a Joint Working Team, involving scientific partners and local stakeholders facilitates the communication between actors. It is possible to conclude from the experiences conducted so far that workshops involving experts and stakeholders facilitate the exchange of information about water stress mitigation options resulting in an increase of stakeholders' knowledge about these issues.

Harmoni-COP is aimed at supporting social learning based on a dialogue and information exchange between stakeholders and decision makers. To this aim, several methodologies and tools to support stakeholders' participation have been applied in different Case Studies. The results have been analysed and included in a handbook to support public participation exercises.

The lesson learnt from the review of these projects is that three main approaches can be adopted to facilitate information exchange among actors, i.e. institutional/legal approach, technical approach and stakeholder approach. The first one is mainly based on creating protocols for data collections and management starting from the analysis of legislations concerning monitoring and data management in the different countries. The technical approach is based on the development of ICT tool, mainly web-base, to facilitate data input and accessibility. The stakeholders approach is based on the negotiation between stakeholders in order to build a shared understanding about information needs.

8.2.3 Integration of different types of information

As discussed in the introductory section, IWRM requires different types of information, ranging from natural sciences data to socio-economic information. The different types of information need to be analysed and integrated so that policy and decision-makers can get an understanding of how the whole system functions. To support IWRM, monitoring programs must be based on the collection of a wide range of information, coming from different sources, to be integrated in order to get as clear a picture as possible of the state of the system.

As discussed in the previous section, one of the goals of the **MANTRA – East** project is the involvement of stakeholders in the information needs elicitation process. From this perspective, the project aims to examine the type and nature of environmental information that is considered important for the decision making, i.e. physical, chemical, ecological, sociological, and economic information. Furthermore, the project aims to define the way in which the different sources of information are used and integrated. From the stakeholders workshop it is concluded that, in order to facilitate the use of information, the cause-effects relationships should be made clear to information users. That is, the information collected by the monitoring system should be organised to make explicit the cause-effects relationships. For this purpose, the DPSIR framework is useful to structure the information concerning driving forces, pressures, state, impact and responses. The decision makers involved in this project seem particularly interested in information about driving forces and possible effects of responses taken.

The **Tisza River Basin** project mainly aims to define an integrated tool to support water management. This tool is based on the integration between different modules, each of them is focused on a well defined environmental problem. The modules are based on the application of models developed considering the requirements of the decision makers and the availability of data. The central modules contain the hydrological and hydraulic models, forming the basis for all other models. The other models are: Catchments models focussing on the control of non-point source pollution; River water quality models for nutrients, oxygen household processes and specific other pollutants that might prove to be necessary when issues are fully identified; Pollution spill models, to form the basis of the establishment of an operative quasi-on-line pollutant wave propagation forecasting system; Lake (wetland) and reservoir ecosystem models to help managing the unique aquatic ecosystems. All these models are based on the integration of different types of information. Moreover, all these models have been integrated in the Tisza River Basin Information System (TRIS), which is mainly based on the use of web-GIS technologies to support the information accessibility and diffusion. From this project is concluded that the definition of an integrated tool has great potential to support the management of water and environmental issues even in such a large river basin like the Tisza. Nevertheless, this can only be assured if the integrated database and the tools developed can be maintained and upgraded for follow-up applications.

The adoption of an integrated perspective in water management is the aim of the **Harmoni-CA** project, dealing with:

- The integration of different physical domains in water management such as hydraulics and ecology;
- The integration of socio-economics and physics, for example the integration of economic water demand and water availability;
- The integration of stakeholder-based approaches.

As regards monitoring activities, one of the tasks of Harmoni-CA is to enhance integration between monitoring systems and modelling, emphasising the role of information and data needs. The key topics are:

- Data availability and accessibility for modelling purposes in the process of implementing the Water Framework Directive;
- Data uncertainty and data assimilation techniques;
- Combined monitoring and modelling.

From the experiences made in this project is concluded that the integration between modelling and monitoring is fundamental to better understand the relationships between the different components of the environmental systems. The use of models allows overcoming the difficulties in understanding the dynamic feedback of the systems, which is made particularly difficult in an environmental context because they are confounded by many factors. Therefore, considering the results of models, information coming from monitoring system could be organised according to dynamic feedback relations. Conversely, the availability of new data permits the revision and updating of models.

The definition of new approaches to the integration of diverse multi-sectorial and multi-disciplinary knowledge in identifying water stress mitigation options is one of the objectives of the **AquaStress** project. AquaStress explores new interfaces between technologies and social approaches, disciplines and sectors, and develops a radically innovative water stress mitigation conceptual framework and guidelines to implement integrated options. It attempts to exploit knowledge generated from traditional disciplines and to show how such knowledge can be integrated in a way which makes it accessible to decision making. The potentialities of Information and Communication Technologies (ICT) to support the integration of knowledge, information and data are tested in AquaStress. The information is assembled in an Integrated Solutions Supporting System (ISSS), which consists of contextually sensitive ICT frameworks to store, recall and present data, information and knowledge. The knowledge stored in the ISSS can be integrated taking into account the stakeholders' perception of local environmental conditions. AquaStress has adopted a stakeholder-driven approach to ensure the success of water mitigation strategies in terms of suitability, acceptance and deployment by end-users. A feature of this approach is an iterative process whereby options are developed and improved through a dialogue between experts and stakeholders. The involvement of stakeholders is considered as crucial both in the definition of water stress mitigation strategies and in their evaluation. In one of the case study, a workshop has been organised in order to define a conceptual model using the stakeholders' knowledge. This model is used to define the cause-effects relationships between the factors influencing the water management in the region. These relationships will be used to structure the information collected by the monitoring system.

Many other research projects deal with the integration of different sources of information in environmental monitoring. Among them, those concerning the integration between scientific and local knowledge seem particularly interesting for the development of a monitoring information system to support IWRM even in data-poor regions. The project "**Combining systematic and participatory approaches for developing and promoting strategies for sustainable land and water management**" can give interesting hints, although it focuses on land resources (Oudwater *et al.* 2003). In fact, it deals with the integration of local and scientific knowledge for environmental management in semi-arid regions, using a GIS-based information system as the integration platform. The project is based on the direct involvement of local community members in providing their knowledge and perception about land resources. Participatory rural appraisal methodology has been used to collect local knowledge, leading to the development of an indigenous GIS map. Furthermore, the local scientific community has been involved in providing its knowledge. A further step concerns the reclassification of land resources combining scientific and indigenous classifications using the data assembled on the GIS. This project is really interesting since it highlights the role that local knowledge can play in supporting environmental monitoring in data poor regions. It also emphasise the importance to adopt well define methodologies to collect and structure local knowledge, to make it reliable for the decision making process.

Harmoni-RIB project deals with the integration of hydrological, ecological and socio-economic information into a DSS for river basin management (Klauer *et al.* 2006). Different phases of the decision process have been identified: framing the decision-making problem; impact analysis; final evaluation, decision support and decision-making. Each of these phases is influenced by the uncertainty related to the collected information. During the problem-framing phase, uncertainty could influence the first analysis of pressures and impacts, and the identifications of potential measures and alternatives actions. Moreover, the uncertainty of information influences the evaluation of possible impacts of actions and alternatives, and, therefore, the final decision. From the experience in different case studies, it was concluded that to support decision making in a complex domain like river basin management the integration of information may be not enough. It is important to take into account and integrate also the uncertainty related to the information used in each step of the decision making process.

The integration between different information to support water management is one of the objectives of **NeWater** project. This project aims to define an Adaptive Monitoring Information System (AMIS) that can be considered part of the learning process, which characterises Adaptive Management. Among the properties of the AMIS, the capability to integrate different sources of information is important to support water management. The architecture of the AMIS has been defined. Several databases should be used to store different types of information. The databases should be used to store data in various formats, such as geographical vector data, data from the monitoring network (time series), and additional information such as qualitative data, reports, pictures, models results, etc. The core of the database should be a module able to combine different information types. Integrating different information types involves dealing with uncertainties. Depending on the source properties, different characteristics of immanent data uncertainties should be distinguished. Some data are subject to technical measurement errors, others to generalisation and interpolation or extrapolation methods, and qualitative information is more or less based on subjectivity. Thus, data based on model results are subject to data uncertainties as well as uncertainties corresponding to the model characteristics. The AMIS has to deal with these different types of uncertainties and produce information about the uncertainties of the data that were used to produce information. Uncertainty management is particularly important when using qualitative information, such as local knowledge. To deal with issue, AMIS is provided by an expert system able to validate the knowledge coming from the local communities.

The integration among different information and data is considered fundamental to deal with complexity in water management. The review of the EU funded suggests different approaches. Some of them are based on the stakeholders' involvement to identify cause-effects relationships between factors. These relationships are used to structure the information collected by the monitoring system. Other approaches are based on the integration between models dealing with different components of the environmental system. However, the potentialities of ICT tools to support integration among information have been deeply investigated and tested. It is important to note, moreover, that the integration of different information to support river basin management implies the integration of the related uncertainties.

8.2.4 Information accessibility and comprehensibility

The importance of environmental information has increased along with the necessity to ensure an appropriate access to the information concerning the environment to all the parties interested in decision-making. Access to environmental data and information is crucial to support public participation. However, these data are not always available to the public and/or easily understandable by all the different stakeholders.

In the **HarmoniCOP** project, the role of information and information tools in supporting participatory river basin management has been investigated (Maurel 2003). The project aimed at developing and providing a methodology to further analyse the application of Information and Communication Technology Tools (ICT), and at assessing their real impact on the improvement of public participation. In fact, ICT has great potential in facilitating information access and communication, but it has also serious risks, such as the absence of metadata on the origin and accuracy of data and "digital-divide", that is, inequalities in information access because many people still are not familiar with information technologies. An analysis of communication patterns to exchange information / knowledge between experts who design ICT tools and people who need relevant information for their part of the decision-making process was carried out. The ICT tools have been clustered in different groups: tools to obtain information (questionnaire); tools for static representation of reality (maps, geographic information system, etc.); tools for dynamic representations of reality, simulations (scenario tools, simulation tools, integrated assessment models, etc.); interactive tools (group support system, form communities, web mapping, etc.). The ICT tools have been evaluated according to the following criteria:

- Communication direction (top-down, bottom-up, bi-directional communication);
- Public size (small working group, general public);
- Usage purpose (management of information and knowledge, perspective elicitation, interaction support, simulation);
- Phases in the public participation process (starting organisation, actors and context analysis, diagnosis of the situation, search for solution).

The analysis has been carried out in different case studies. From this analysis it could be drawn that the ICT can be really important to support information accessibility and sharing, in order to facilitate the social learning process in river basin management. Nevertheless, the usability of a tool might be influenced by economic, technical, institutional and cultural factors. Moreover, the sharing of informational sources and co-design of IC tools plays an important role. In fact, the stakeholders involvement in a participatory process has to be based an equal access to informational resources. ICT can facilitate the comprehensibility of information through visualisation techniques. Participating in the co-design of an informational tool facilitates the acknowledgement of both expert and local knowledge and offers a positive context for bidirectional communication and mutual understanding.

The **AquaStress** project is strongly oriented towards the adoption of a stakeholder-based approach for the identification of water stress mitigation strategies. Thus, the accessibility of knowledge and information about mitigation options is crucial for the success of the project. One of the aims of AquaStress is the development of a web-based system able both to provide guidelines to the stakeholders concerning how to deal with water stress problems and to provide "intelligent" support to the learning process will be developed. This system is able to support the behavioural changes of each stakeholder and to facilitate information and knowledge exchange among stakeholders in order to support the collective learning process. The system supports individual learning facilitating the access to the knowledge collected in AquaStress case studies. An user-friendly interface has been developed able to facilitate the knowledge retrieval according to several parameters, e.g. the characteristics of the sites and of the water management problems.

In the **MANTRA-East** project, to enhance information diffusion, the main "information channels" in the different countries have been identified. From the results of this analysis it is concluded that internet cannot be always considered as the best way to share information because in some countries most of the population is not familiar with ICT. The lack of internet access for the majority of the general public shift the focus on traditional information diffusion means, like:

- community information sessions and campaigns, to inform the general public on environmental issues and to provide a medium for suggestions, comments and participation;
- production and distribution of written information;
- citizen juries, formed by demographically representative panel of citizens, to discuss the issues surrounding a particular policy area.

These approaches have been suggested particularly to disseminate information concerned with local water related issues at the local level. At the end of the project, the Lake Peipsi web portal has been developed to collect, process and disseminate information pertaining to the Lake Peipsi region. The portal facilitates the research process by allowing users to find the most relevant information, and it supports users to understand the extent of connections between different concepts and keywords encountered on a daily basis.

The enhancement of data comprehensibility is an implicit aim of the **Harmoni-RIB** project. Harmoni-RIB focuses on the definition of data management methodologies to support decision-makers in a decision context characterised by a high level of uncertainty. To this aim, a preliminary review of current practice with respect to representative river basin data sets and the way they can be accessed by users has been carried out. At the same time, the requirements for uncertainty information on input data and model results in relation to water management according to the WFD were identified. An overview of the data demand and other data-related issues for the Water Framework Directive (WFD) is provided in one of the project deliverables. From the analysis it is concluded that to increase the comprehensibility of data to be used in a decision process, it is important to provide explicit information about the related uncertainties. For this reasons Harmoni-Rib aims to establish a practical methodology and a set of tools for assessing and describing uncertainty originating from data and models used in decision-making processes for the production of integrated water management plans, and to provide a conceptual model for data management that can handle uncertain data and implementing it for a network of representative river basins.

Given the increasing importance of stakeholders' participation in river basin management, facilitation of access to data and information is considered fundamental. Several EU funded projects deal with this issue. Most of them are focused on the development of IC-tool to facilitate data and information sharing. The possibility to define different user interface to support the access to environmental data and information also for non expert system has been deeply investigated. Nevertheless, in many countries a large part of the population is not familiar with ICT. Thus, it is fundamental to adopt the most suitable "information channel" according to the specificities of the considered area. Moreover, the tool used to facilitate information access could be different according to the type and scale of information.

8.3 CONCLUSIONS – WHAT HAVE WE LEARNED? WHAT IS MISSING?

The properties of a monitoring information system able to support water management dealing with the increasing complexity are described in this chapter. Starting from IWRM and AM requirements for monitoring systems, a review of EU funded research projects was carried out that aimed at verifying whether, and if so, to what extent, research in the monitoring field is moving towards these goals. Four main topics have been considered, i.e. interconnection between monitoring/information management and decision processes; information exchange among actors; integration of different types of information; degree of information accessibility for both the decision-makers and stakeholders.

The most significant changes in monitoring and information management can be found in the enhancement of information accessibility through the use of Information and Communication Technology (ICT) to support visualisation and communication. In the projects considered, ICT is imperative in providing timely, accurate and understandable environmental information. The spatial nature of environmental data has favoured the use of Geographic Information Systems (GIS) and associated technologies. GIS is used for data exploration and visualisation, and improving communication among the decision-makers. Moreover, various attempts have been made to improve the user interface of these tools in order to allow non-expert people to make up their own information.

In most of the reviewed projects identification of users' requirements was used as the basis for interface development. Nevertheless, while ICT is widely considered as a useful tool to enhance democratisation of water management decision process, as more people have access to more information, it is important to care about the "digital divide". Use of ICT could exclude a large part of the population from the access to information and, therefore, from the decision process. Ample attention is therefore needed to integrate between traditional and innovative means of diffusing environmental information.

The development of a framework to support transboundary water management is fundamental, since many multi-national river basins are present in Europe. Moreover, the WFD requires the creation of water management units according to river basin boarders, not administrative boarders. Therefore, several EU research projects aim at the enhancement of information exchange among different actors at different levels and working in different national legal frameworks. Most of them aim at defining guidelines in monitoring data collection and information management to facilitate information exchange because the differences in legal framework and in data collecting methodologies represent strong barriers to information exchange. The definition of such guidelines succeeded up to a certain point, but still institutional changes are required. This is particularly true in transboundary river basin at the European fringe, where the obligations for EU countries and non-EU countries are different. Therefore the definition of strategies to facilitate data collection, analysis and exchange needs to be based on a long debate and negotiation process between all the involved countries.

Integration between monitoring processes and decision processes has been recognised as an important issue by only a few EU projects. This conclusion can be explained by considering that most of the projects start off from an already well-defined environmental problem and focus on the use of modelling tools to support environment management. The monitoring system is considered only as a tool to collect the necessary information. There are no feedbacks mechanisms between monitoring and the decision process. An iterative perspective of monitoring design is therefore missing in most projects. In fact, only in the Mantra-East project monitoring design is considered as an ongoing activity, which iteratively redefines information needs, data collection and data interpretation, according to changes in the policy phase. From the review it is concluded that monitoring is not sufficiently linked to the policy making process. Therefore, research aiming to develop methodologies for monitoring design based on a strong and iterative relationship between monitoring systems and decision processes is much needed.

In some of the projects in this review, there is no involvement of different parties in monitoring design. The monitoring design process should involve all the interested parties, ranging from scientists to local communities, in order to exchange mental models as the basis for information needs elicitation. The Mantra-East, AquaStress, NeWater and Tisza River Basin projects are good examples of involvement of stakeholders in the design of monitoring and evaluation systems.

Most of the considered research projects adopt an integrated perspective in data/information collection and management, trying to reproduce the complex system of relationships according to a systemic view of the environment. Therefore, in most of the projects information and knowledge concerning different elements (i.e. water quality, water quantity, biodiversity, land use, climate, etc.) are considered and integrated into models (integrated models) and/or information management systems. Nevertheless, although integration of information is considered fundamental in most of the projects reviewed, only few of them deal with the design of monitoring systems that are able to integrate different types of information sources. In particular, integration between local and scientific knowledge is still far from becoming an important research issue in environmental monitoring. Further research is needed to overcome the limitations hampering the usability of local knowledge, i.e. data credibility, data comparability, spatial scale, etc. The validation of local knowledge using an Expert System is one of the aims of NeWater project.

Concluding, from the review of EU projects we have learned that different approaches can be adopted to deal with the issues related to monitoring and IWRM. Two main groups can be defined, i.e. technological-driven approaches and stakeholder-driven approaches. The former are mainly based on the development of new technologies, and/or improvement of existing ones, to facilitate data and information integration, to enhance data accessibility and to improve the exchange of data between actors. The latter are based on the direct involvement of stakeholders in the design of information production and management tools. These approaches are not mutually exclusive, but should be integrated, leading to the co-design and co-development of tools that can become particularly effective.

Nevertheless, only few projects start from the premises that the monitoring system has to become part of the learning process characterising river basin management under conditions of complexity. A deeper change of perspective is needed. Monitoring and decision-making processes cannot be considered in parallel, but there is a need for a strong and continuous interrelationship. The monitoring program needs to be adapted to the different stages of the policy definition process because each stage requires other types of information. Where current monitoring is adapted after changes in policy or management are implemented, adaptive water monitoring should be an integral part of the water management process as well as the water governance process. This cannot be done without the establishment of an iterative dialogue between information users and producers. To become part of the whole learning process in environmental management, monitoring system should be adaptive itself, allowing changes in information needs and monitoring strategies according to the results of learning in both water management and information production and management.

8.4 REFERENCES

Bender, S., Mieseler T. and T. Rubbert. (2004) *TransCat Deliverable 4.1 - Manual - Standard Procedure for Data Acquisition*. URL: http://www.transcat-project.net/.

Checkland, P. (1981) *System Thinking, System Practice*. Wiley, Chichester, UK.

Checkland, P. (2001) Soft System Methodology. In Rosenhead, J., Mingers J. (eds) *Rational Analysis for a Problematic World*. John Wiley and Sons, Chichester, UK.

Cofino, W.P. (1994) Quality management of monitoring program. *Proceeding of the First MTM Conference* - International workshop on monitoring and assessment in water management. The Netherlands.

Denisov, N. and L. Christoffersen (2001) *Impact of environmental information on decision making processes and the environment*. UNEP/GRID-Arendal, Occasional paper 01. Arendal, Norway. http://www.grida.no/impact/.

Enderlein, R.E. (2004) Role and Sources of Information. In Timmerman, J.G., Willem, H., Behrens, W., Bernardini, F., Daler, D., Ross, P., van Ruiten, K.J.M., Ward, R.C. Lelystad (eds). *Information to support sustainable water management: from local to global levels*. The Netherlands, RIZA.

Fazey, I., Fazey, J.A., and Fazey, D.M.A. (2005) Learning More Effectively from Experience. *Ecology and Society*. 10(2), 4. [online] URL: http://www.ecologyandsociety.org/vol10/iss2/art4/

Giordano, R. and Vurro, M. (2005) *Critical review on existing monitoring systems*. NW1.6.1. URL: http://www.newater.info.

Giordano, R., Haase, D., Liersch, S., Timmerman, J.S. and Vurro, M. (Forthcoming) The role of information and new concepts for adaptive monitoring systems. Paper submitted to *Ecology and Society*.

Gouveia, C., Fonseca, A., Camara, A. and Ferreira, F. (2004) Promoting the use of environmental data collected by concerned citizens through information and communication technologies. *Journal of Environmental Management*. 71, 135-154.

Haklay, M.E. (2003) Public access to environmental information: past, present and future. *Computers, Environment and Urban Systems*. 27, 163-180.

Kamphorst, E., Jørgensen, L. F., Van Griensven, A. and Vanrolleghem, P.P. (Eds.) (2005) *Joint use of modelling and monitoring for implementing the Water Framework Directive* – Workshop Report. April 4-5th 2005, Ghent, Belgium

Keenan, S.L., Rex Hartson, H., Kafura, D.G. and Schulman, R.S. (1999) The Usability of Problem Taxonomy: A Framework for Classification and Analysis. *Empirical Software Engineering*. 4, 71 – 104.

Klauer, B., Dunbar, M., Brown, J., and Petersen, T. (2006) *Concept for integrating uncertain hydrological, ecological and socio-economic information into a decision support system for river basin management*. URL: www.harmonirib.com.

Kolkman, M.J., Kok, M. and van der Veen, A. (2005) Mental model mapping as a new tool to analyse the use of information in decision-making in integrated water management. *Physics and Chemistry of the Earth*. 30 (4-5), 317-332.

Jolankai, G. (Ed.) (2004) *The Tisza River Project – Final Report*. URL: http://www.tiszariver.com.

Lane, D.C. and Oliva, R. (1998) The greater whole: Towards a synthesis of system dynamics and soft system methodology. *European Journal of Operational Research*. 107, 214-235.

Lessard, J. (1998) An Adaptive approach to planning and decision making. *Landscape and Urban Planning*. 40, 81-87.

Maurel, P. (Ed.) (2003) *Public Participation and the European Water Framework Directive – Role of Information and Communication Tools*. Cemagref, Montpellier. URL: www.harmonicop.info

McDaniels, T L, and Gregory, R. (2004) Learning as an objective within a structured risk management decision process. *Environmental Science & Technology*. 38 (7), 1921-1926.

Oudwater, N. and Martin, A. (2003) Methods and issues in exploring local knowledge of soils. *Geoderma*. 111, 387-401.

Pahl-Wostl, C. (2007) The implication of complexity for integrated resources management. *Environmental Modelling and Software*. 22, 561-569.

Raiffa, H., Richardson, J. and Metcalfe, D. (2002) *Negotiation Analysis*, Harvard University Press, Cambridge, Massachusetts.

Ringold, P.L., Mulder, B., Alegria, J. Czaplewski, R.L., Tolle, T. and Burnett, K. (1999) Establishing a regional monitoring strategy: the Pacific Northwest Forest Plan. US Environmental Protection Agency, Corvallis, Oregon, USA. *Environmental Management*. 23(2), 179-192.

Roll, G. and Timmerman, J.G. (2006) Communicating with stakeholders and the public. In Gooch, G.D. and Stalnacke, P. (eds). *Integrated Transboundary Water Management in Theory and Practice: Experiences from the new EU Eastern Borders*. IWA Publishing, London, UK.

Rosenhead, J., and Mingers, J. (2001) A New Paradigm for Analysis. In Rosenhead, J., Mingers J. (Eds.) *Rational Analysis for a Problematic World*. John Wiley and Sons, Chichester, UK.

Smit, A.M. (2003) *Adaptive monitoring: an overview*. DOC Science Internal Series, vol. 138. Department of Conservation, Wellington. 16 p.

Sterman, J.D. (2000) *Business dynamics – systems thinking and modeling for a complex world*. MacGraw Hill, Boston, Massachusetts, USA.

Timmerman, J.G. (Forthcoming) *Bridging the water information gap. A framework for linking environmental information to water management policy through specification of information needs*. PhD Thesis, Wageningen University, Netherlands.

Timmerman, J.G. and Langaas, S. (2004) Conclusions. Pages 240-246 in: Timmerman, J.G. and Langaas, S. (Eds.) *Environmental information in European transboundary water management*. IWA Publishing, London, UK. ISBN: 1 84339 038 8.

Timmerman, J.G., Ottens, J.J. and Ward, R.C. (2000) The information cycle as a framework for defining information goals for water-quality monitoring. *Environmental Management*. 25(3), 229-239.

Ward, R.C. (1986) Framework for designing water quality information systems. Pages 89-98 in: Lerner, D. (Ed.) *Monitoring to detect changes in water quality series: proceedings of the Budapest symposium*, July 1986. International Association of hydrological Sciences, Wallingford, UK.

Ward, R.C. (1994) Monitoring tailor-made: what do you want to know? *Proceeding of the First MTM Conference* - International workshop on monitoring and assessment in water management. The Netherlands.

9

Integration of spatial planning in flood management in European research

Hans de Moel and Jeroen Aerts

9.1 INTRODUCTION

Traditionally, water management in the EU is a relatively autonomous sector focused on developing technical solutions to control both water quantity and quality (Pahl-Wostl *et al.* 2006). However, increased population growth and economic developments are increasingly pressing water resources and their management (GWP-TAC 2000). Managing water availability, water quality and flood impacts is therefore becoming progressively more complex. Therefore, a transition to a more integrated approach to manage water resources, by incorporating different perspectives (human & nature, land & water, coast & catchment etc.), has been introduced known as integrated water resources management (IWRM; see GWP-TAC 2000). In IWRM, ideally different actors and sectors cooperate in a participatory system to achieve a common objective to manage water resources in a sustainable way. IWRM has been adopted in the Water Framework Directive (WFD) by the European Parliament and Council in 2000 (European Commission 2000), but this directive mainly concerns water quality and availability.

More recently, the EU has also recognised the need for such an integrated approach in flood management and a flood directive is almost agreed that will compel member states to make flood assessments, risk maps and flood management plans (CEC 2006). The motivation behind this directive are the rising impacts of floods during the last decades, increasing potential damage of floods in the near future and the notion that severe floods events might become more common due to global climate change (Milly *et al.* 2002). The integration of spatial planning and water management is considered to be of special importance with respect to integrated flood management (Burby *et al.* 2000; Hooijer *et al.* 2004), as they are the main sectors managing the occurrence and the impact of floods.

In this chapter we will explore how spatial planning is integrated in flood risk management projects in EU research projects, how their results contribute to the integration of flood management and spatial planning in practice and what gaps still exist in EU research that should be addressed in the future.

9.2 FLOOD RISKS

Floods account for about a third of all economic losses insured worldwide and about half of the people that were killed by natural hazards were casualties of floods (Munich RE 2005). Floods are therefore regarded as the most dangerous natural hazard (ARMONIA Project 2005c). Moreover, because of the high economic losses associated with floods, there is a rich history of flood management practices aiming at preventing floods from happening and reducing costs associated with them (known as risk management). Before going further into flood risk management, some concepts need to be clear though as their use is not always consistent.

Flooding, which is in essence the covering of land by water, often refers only to fluvial floods, from rivers. It should be noted, though, that there are other are many different sorts of flood types (see e.g. Kron 2002):

- Floods from high river water levels, called: river flood, fluvial flood, alluvial flood
- Floods due to heavy rainfall, called: inundations, flash flood, off-floodplain flood, meteorological flood
- Floods from the sea, called: tidal flood, storm surge
- Floods due to full sewers: urban flood, drainage overflow
- Floods due to (ice) dam breaks, groundwater rise, waterlogging, backwater, tsunami and more

Furthermore, *Risk* is also a term used in many different ways. In some studies, risk is defined as the probability a disaster occurs (e.g. the U.K. Environment Agency[1]). However, in most studies risk includes potential damage as well (e.g. CEC 2006; Vis *et al.* 2003). Potential damage has in turn a rather broad meaning in itself and may incorporate financial damage (cost of replacement), economic damage (business interruption), ecological damage (loss of biodiversity/habitats), cultural damage (monuments), human damage (loss of life, health) etc. In general terms, the scientific community defines risk as the probability of an event multiplied with its consequences (Kron 2002). A detailed definition has been proposed by Kron (2002), using three components to determine risk (Figure 1), which is also used by, or been adopted by various other groups (Lavalle *et al.* 2005; Samuels and Gouldby 2005; Schmidt-Thomé *et al.* 2006). In this definition the terms are defined as follows:

- Hazard: the probability and magnitude of a specific event (e.g. flood).
- Exposure: the value of a specific area: including population, land-use, infrastructure, economical and ecological assets etc.
- Vulnerability: resistance or coping capacity to a disaster: including protection measures, GDP/capita, insurance availability etc.

As displayed in Chapter 6, there is not a univocal definition of the term 'vulnerability'. Though broadly defined as 'the potential to be harmed', vulnerability is also used as exposure by some researchers (e.g. the German Centre for Disaster Management and Risk Reduction Technology[2]). For the FLOODsite project a comprehensive glossary of definitions has been drawn up (Samuels and Gouldby 2005), covering the broad range of terms in the field of risk management and reflecting the most of the consensus among the experts in this field.

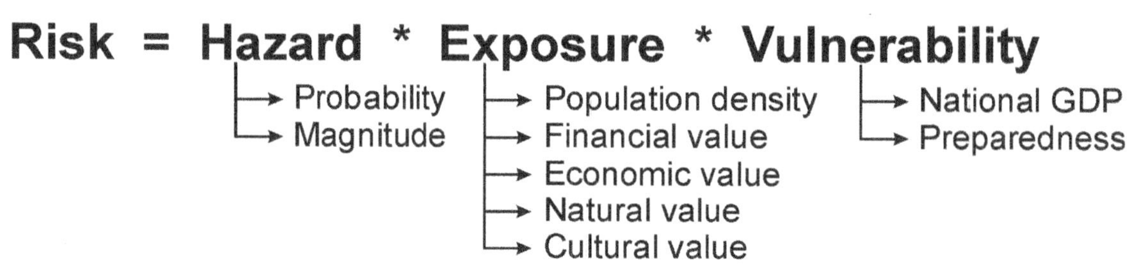

Figure 1: Definition of the concept of risk following Kron (2002)

During the last two decades, economic losses resulting from large floods have increased significantly (Figure 2). This is a direct result of more people and industry settling in floodplain areas without taking into account long term changes and possible risks associated with such developments (ARMONIA Project 2005a; Burby *et al.* 2000; Godber 2002). Because of these developments, exposure increases and with it flood risks. As floodplain developments are likely to continue in the future, flood risks are likely to increase as well. In addition, the probability and magnitude of flood events is increasing as global warming enhances the hydrological cycle, which results in more severe rainfall events (IPCC 2001; Milly *et al.* 2002) and consequently an increase in the hazard and thus risk. Because of the increasing risk, traditional flood management using optimised engineering solutions is becoming difficult. Long term changes in flood risks are inherently unpredictable. Hence a optimised design is probably not robust enough to cope with unexpected events. This increased awareness about the problem of floods and its uncertainties have stimulated many authorities to use a more holistic and flexible approach by managing flood risks, such as disaster and risk management approaches or adaptive management (e.g. Germany: Petrow *et al.* 2006; UK: Tunstall *et al.* 2004; Netherlands: Vis *et al.* 2003).

---

[1] http://www.environment-agency.gov.uk/subjects/flood/1217883/?version=1&lang=_e

[2] www.cedim.de

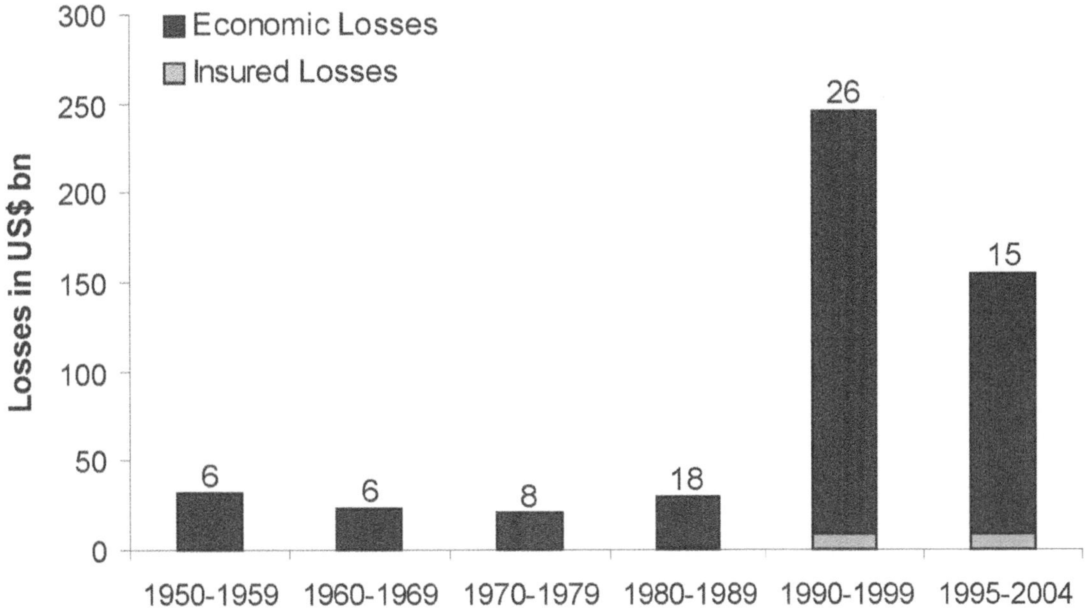

Figure 2: Economic and insured losses for 'great' floods worldwide from 1950 to 2004. Floods are classified as 'great' when interregional or international assistance is necessary. The numbers above bars represent the amount of 'great' floods during that specific period. Data adopted from Munich Re (Munich RE 2005).

9.3 MANAGING FLOOD RISKS

Managing disasters like floods can be structured according to the so-called disaster management cycle (see e.g. WHO 2002), which illustrates the ongoing process of how different actors involved try to cope with a disaster (Figure 3). The disaster management cycle (DMC) consists of five components, all aimed to reduce one or more of the different terms of the risk definition.

(1) *Preventive* measures aim to reduce the 'hazard' potential of floods by lowering the probability and/or magnitude of floods. They include mainly structural measures like dikes and diverting channels, but also spatial measures like retention areas and floodplain widening/lowering can be used to lower the hazard potential.

(2) *Mitigation* efforts intent to minimise the effect of a flood, mainly by reducing the 'exposure' of an area. For instance, sound land-use regulations can influence the potential damage and flood resilient building helps to reduce the potential damage. Additionally, the availability of flood insurance helps to mitigate the impact of a flood by reducing the 'vulnerability' of a community.

(3) *Preparedness* for a flood reduces the 'vulnerability' and can be increased by early warning systems and detailed emergency plans.

(4) In *response* to a flood, the 'vulnerability' to a flood can be lowered by providing emergency relief and executing rescue operations.

(5) During the *recovery* phase, after an event has taken place, the 'vulnerability' of a community is dependant on aspects like medical care, rebuilding grants and so on.

While the components of the disaster management cycle are more or less chronologic, it is not a static path. Especially the 'before event' components do not necessarily happen in this particular order. Measures related to flood risk management can be classified by this framework and naturally some measures have more effect than others, where the most frequently used is not always most effective. For instance, the role of public awareness is greatly underused within flood management (Siegrist and Gutscher 2006), but can have substantial effects, as can be illustrated by the 1993 and 1995 floods in Germany and the Netherlands. While of comparable magnitude, damages were significantly lower in 1995 than in 1993 (50% and 35% lower in Germany and the Netherlands respectively), mainly due to lower household good damages as the general public had learned from their previous experience (Kron and Thumerer 2002; Wind *et al.* 1999). The use of spatial planning measures to reduce losses from natural hazards is considered one of the most promising approaches (Burby *et al.* 2000; Hooijer *et al.* 2004). For example, it has been demonstrated that local comprehensive plans considering natural hazards could have reduced insured losses by about 1%, which would have lowered the insured losses in the USA by $213 million (Burby 2005).

Figure 3: Illustration of the 'Disaster Management Cycle', coupled to the different terms of the definition of risk, forming a framework for risk management strategies.

Spatial planning measures can take place in all of the 'before event' components and can act on each of the risk terms. Most spatial planning measures are designed to *mitigate* flood risks. Regulating land use in flood prone areas is the most apparent one but also the establishment of flood resilient building codes in certain areas lowers the exposure. There are however also many possibilities to reduce the flood hazard by *preventive* spatial measures. The likelihood and magnitude of a flood can be reduced by reserving areas for flood plain widening, for the removal of bottlenecks or by appointing retention areas. Lastly, spatial planning can theoretically play a role in the *preparation* for a flood event as well, for instance by optimising infrastructure to facilitate evacuation routes from high risk areas.

Important in the effective contribution of spatial planning to flood management is to distinguish the different tasks of spatial planning levels, ranging from European to municipality scale (Figure 4). Most often spatial planning is associated with nation-wide actions and regulations. At the national level certain goals on spatial development are distinguished and frameworks are created to meet those goals at the lower scales. At the regional level, plans are drawn up for spatial or physical structures and development, linking the overarching state-wide perspective to the local level. It is the local level where concrete decisions on land uses are taken and policies are created to guide land and resource use inside its administrative borders (ARMONIA Project 2005b).

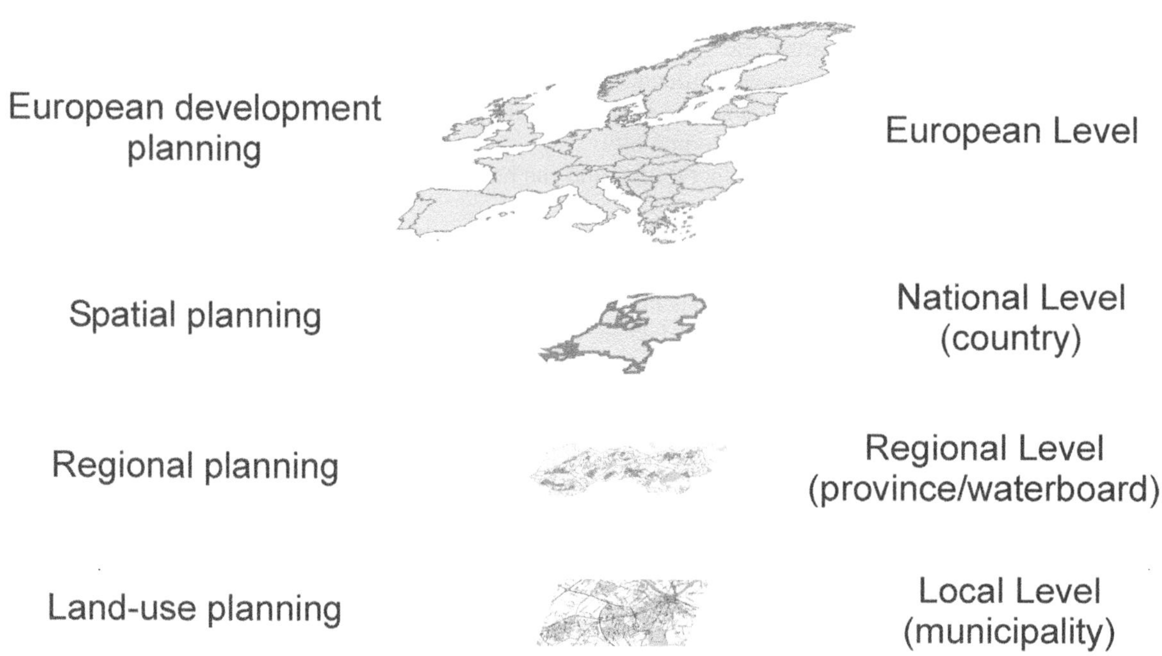

Figure 4: Different levels of spatial planning and their corresponding terminology. Largely taken after Greiving *et al.* in the annex of deliverable 3.1 of the ARMONIA project (ARMONIA Project 2005b).

Evidently, spatial planning can play an important role with respect to reducing flood risks. However, additional information about the flood hazard is required from water managers and close cooperation with the water management sector is thus required. Conversely, spatial planners are in the position to supply data about exposure (developments), allowing water managers to purposefully plan new measures. Clearly, an integrated approach in water management, such as the IWRM approach, has a profound added value with respect to flood risk management and will increase the adaptive capacity against floods.

9.4 REVIEW OF EU RESEARCH

Integrating autonomous sectors is a long process and a substantial amount of relevant research needs to be conducted before any real integration can occur. In this section selected research projects financed by the EU addressing flood risks are reviewed using the disaster management cycle and specific attention is paid to the role of spatial planning in these projects. This review should not be considered comprehensive. EU Projects related to flood management have been selected from a pre-selection made by the NeWater team (Blind *et al.* 2005) and various recent INTERREG projects on flood management have been added. In total 10 projects have been identified in which both flood risk as well as spatial planning issues have been addressed. Table 1 gives an overview of the scale and nature of these projects and indicates on which parts of the disaster management cycle the project focuses and what kind of actions related to spatial planning are addressed in the project. Finally, the projects have been classified in 5 classes, according to how far the integration of spatial planning and flood risk management has been put into practice. The first class forms projects generating supportive information for a possible integrated approach (like mapping and modelling projects). The next two classes cover assessments of how far spatial planning is currently integrated in flood management and assessments of the effectiveness of certain spatial measures. Finally, the last two classes encompass projects that actively work to integrate spatial planning and flood risk management or put integrated measures into practice.

9.4.1 Projects overview

The research projects of ESPON (Schmidt-Thomé *et al.* 2006) and the Joint Research Centre (Lavalle *et al.* 2005) both involved the mapping of fluvial flood risks on a European scale, including estimates on exposure and vulnerability. The aim of these projects was to support EU regional development in the context of European cohesion and solidarity. As such, these projects do not address a specific component of the DMC but rather illustrate in which areas flood management is most urgent. Considerable sacrifices had to be made though with regard to the detail of the data because of the large scope of the projects, especially with regard to the hazard indicator, resulting in substantial differences between the outcome of the two projects. Nevertheless, such mapping projects are typical examples of projects where data from the water sector (flood hazard) and the spatial planning sector (land use) are combined to get a greater insight into a larger issue (flood management) and as such contributes to integrating spatial planning into IWRM, albeit on a very basic level.

Another research project that looks at risk mapping on a European scale is the ARMONIA project. ARMONIA aims to create a harmonised methodology for integrated risk mapping of all natural disasters in order to improve the effectiveness of spatial planning procedures in hazard prone areas of the EU (ARMONIA 2005a). The project is still ongoing but has already finished a thorough survey (ARMONIA 2005b) on the approaches, norms and practices of spatial planning with regard to risk management practices in several European countries (Finland, France, Germany, Greece, Italy, Poland, Spain and the United Kingdom). For the greater part flood risks are addressed by the surveys and besides describing the national frameworks of flood management, examples of best practices are used to illustrate how spatial planning can *mitigate* flood risks and which problems are faced in practice. The analysis of the survey showed that hazard/risk mapping activities are often driven by recent, often frequent, events. As a result the impact of rare events is regularly underestimated. The analysis further showed that, while it is obligatory in most countries to create hazard maps, there is often no obligation to look at exposure and/or vulnerability. As a consequence, risk management strategies are based on hazard information rather than risk information. Spatial planning is only plays a minor role in such risk management strategies. Furthermore, these strategies are generally contrived at the regional level, where they have to be implemented at the local level. Many programs and measures fail at the local level as the recommendations from the regional level are usually not binding and local authorities often prioritise other (short-term) issues (like unemployment or infrastructure development) over flood issues. Finally, the best practice examples show that coordination of all actors is of vital importance, where currently they usually operate separately. Overall, the survey and its corresponding analysis presents valuable insights into the current integration of spatial planning in flood risk management practices in various countries and the best practices yield important lessons that can be used for future integration of spatial planning into IWRM in order to *mitigate* flood risks.

Table 1: Overview of various EU funded projects addressing both spatial planning and flood management. The small crosses indicate that a certain issue is addressed by the project, but is not directly related to spatial planning.

	Status	Flood Type			Scale			Component Disaster Management Cycle					Spatial Planning related activity					Other Activity				Integrative Character Project				
Project		river	coastal	rain	local	catchment	Europe	prevention	mitigation	preparedness	response	recovery	Mapping Flood hazard/risks	Modelling Floods	Survey SP &FRM	Flood Prevention Measures	Flood Mitigation Measures	Exchanging Knowledge	Decision Support System	Emergency Planning Plans	Raising Awareness	Support FRM	Assess current integration	Assess SP measures in FRM	Integrate SP and FRM	Carry out integrated measures
SDF	ongoing	X			X	X		X								X					x					X
ESPACE	ongoing	X	X		X	X	X		X							X	X				x			X	X	
MITCH	finished	X						X	X	x								x							X	
ELLA	finished	X				X		X	X				X			X	X	x				X		X		
IRMA-SPONGE	finished	X				X		X	X	x				X		X	X		x					X		
FLOODSITE	ongoing	X	X		X	X	X	X	X	X	x	x		X		X	X	x	x	x	x			X		
FLOWS	finished	X		X	X				X	x			X	X	X			x	x		x	X	X			
ARMONIA	ongoing	X	X		X	X	X		X				X		X				x			X	X			
EU-JRC	finished	X					X						X									X				
ESPON	finished	X					X						X									X				

Mapping, modelling and assessing flood risks flood risks in the current situation is already quite a task. However, as global climatic change is expected to alter the hydrological cycle and with it the possibility and magnitude of floods, an extra dimension is created. The FLOWS (Floodplain Land Use Optimising Workable Sustainability) program specifically addresses this issue. Like many projects, FLOWS also acknowledges the potential spatial planning has in facing the problem of increasing flood risks in a changing climate. Their final aim was to improve the sustainability of floodplain development, considering the projected changes in climate. For this purpose, over 40 projects have been carried out focusing on risk mapping, modelling, spatial planning, water management and awareness creation. A large focus of the project, as well as their main conclusions focus on awareness creation in a positive way starting at a young age (FLOWS 2006). However, an evaluation of mapping and modelling techniques, necessary tools to assess the effect of spatial measures, has also been carried out (Wetterskip Fryslan 2006). Furthermore, a survey on how the different partner countries integrated flood risk information into their planning decision making process was carried out (Forbes Tuners 2004). While an analysis of the survey lacks, the information provided does give some insight and starting points on how planning is incorporated in flood risk management in these countries.

From studies like the above it becomes increasingly clear that effective flood management includes many different aspects and that measures are not always sector specific. As an integrated approach is clearly needed, the FLOODsite project was initiated. FLOODsite covers the many dimensions of flood management by bringing together a variety of partners which all work on different, complementary, aspects of flood management. This comprehensive project is the only one where all phases of the disaster management cycle are investigated. Over almost the entire range of flood analysis and management aspects current measures are identified and evaluated. Additionally, new innovative measures are developed and evaluated ex-ante. In the end they aim to produce an integrated methodology for flood risk analysis and management in Europe, consistent for floods from rivers, as well as estuaries and the sea. To disseminate this knowledge, a specific part of the project is devoted to distribute the results to professionals, the general public and school children by creating a 'Best Practices Guide' series, web based models and information, and educational material for primary and secondary schools as well as under and post graduates (FLOODsite 2005). FLOODsite is a model example of a project where integrated flood management is explored, encompassing all components of the DMC. Spatial planning is one of the many sectors in flood management that is included and the results have the potential to greatly contribute to our existing knowledge of the use of spatial planning in IWRM, especially its role within the whole of integrated flood risk management.

Not only is flood management a cross-sectoral field, it is also a transnational issue as many catchments are not bound by borders. This became painfully clear after major floods in the Rhine and Meuse basins (1993 and 1995) and the Elbe basin (2002). This triggered two INTERREG projects in these areas: IRMA-SPONGE and ELLA.The Rhine basin already had a long tradition in international cooperation. Therefore the IRMA-SPONGE umbrella program could focus more on the development of sustainable river basin management options for the Rhine and Meuse rivers, also with a strong focus on spatial issues. The wide range of research performed in this project started from a perspective in which spatial planning and flood management successfully cooperate and then looked at the effect of selected *mitigation* and *prevention* measures. For instance, modelling tools were created (van Deursen and Middelkoop 2002) to evaluate the hydrological responses of land-use changes in the catchments and a decision support system was created with which the effects of floodplain widening and detention measures could be investigated (Bons *et al.* 2002). Furthermore the regulating role of spatial planning in flood prone areas (Böhm *et al.* 2002) and ways to combine land use functions as a flood management measure (Vis *et al.* 2002) were explored. Because successful implementation of new measures is only possible when the decision is carried by the different stakeholders, a simulation tool to provide insights into river and floodplain management was created to better understand the integrated nature of the problem and the interests and considerations of others (Heun *et al.* 2002). The main conclusions that were drawn from this program have been summarised by Hooijer et al. (2004) and show that: i) measures upstream (like land use changes or retention areas) are not considered effective with regard to preventing extreme flood events in the lower Rhine and Meuse river; ii) in order to reduce flood risks, more can be achieved by damage limitation (*mitigation)* then by control measures (*prevention*) and iii) flood risk management should become more a matter of spatial planning and land use adaptations near the area at risk. Overall, the research illustrates the added value of IWRM and provides valuable information on which spatial measures are best used in the Rhine catchment to combat flood risks.

In the Elbe catchment international cooperation with regard to river floods was not as developed as in the Rhine basin. The ELLA project therefore aimed to create a transnational network of spatial planning and water management authorities to develop internationally agreed concepts for flood management. As a basis for these plans, hazard maps have been created for the Elbe catchment (Elbe-atlas) and specific areas where action is needed were identified. Additionally, potential *preventive* and *mitigative* measures that could improve the situation have been assessed. By bringing together water authorities and spatial planners ELLA has taken the first step in integrating spatial planning in international flood management and the conducted research on spatial measures can form the basis for future projects or at least prevent future error (ELLA 2003).

Another effort to bring together spatial planners and water managers was initiated by the MITCH program (Mitigation of Climate Induced Natural Hazards). MITCH was set up tot disseminate research results concerning climatic hazards, including floods, to practitioners and other stakeholder. Using a series of workshops and an active web site knowledge was transferred to the hazard planning management community and end users needs were matched with research capabilities and vice versa (MITCH 2007). In this way MITCH tried to contribute to the work floor integration of spatial planning and flood management in order to *prevent*, *mitigate* or *prepare* for floods.

The most far reaching effort with respect to practical integration is perhaps the ESPACE program (European Spatial Planning Adapting to Climate Events). This program is not a traditional research program, but rather a coordinated effort to put climate change on the agenda of policymakers and to raise awareness on the necessity of adapting to climate change and to incorporate adaptation policies into the spatial planning agenda at all levels. Adapting to floods in the light of climate change is one of the main issues in this project. Spatial planning policies are reviewed and new policies will be developed in the project. The end result is to develop a transnational approach to adapt to climate change within spatial planning mechanisms from local to European level (ESPACE 2006). If the final goals are met, the framework of the ESPACE program might serve as a very valuable reference for European governments in their effort to bring together the spatial planning and flood management sector in order to prevent floods and mitigate flood risks.

The SDF (Sustainable Development of Floodplains) is even more focused on end-users. In this project several flood damage prevention measures are physically implemented in an integrated way, balancing it with other spatial functions in the floodplain and by transparent interaction with all stakeholders. While most pilot projects are actual implementation projects, some are feasibility studies. All projects are *prevention* oriented and deal with physical measures in the floodplain. These measures aim to increase the discharge capacity, create additional space for retention of water or combine retention with other floodplain functions like nature, recreation and agriculture. In all these projects public communication and involvement plays an important role in order to raise awareness and increase commitment to flood prevention issues (SDF 2005). Even though the SDF project only involves spatial measures to prevent the flood hazard, it contributes significantly to the integration of spatial planning in IWRM as crucial experience is gained with respect to the actual implementation of spatial measures. Especially the communication between different stakeholders in the decision making process is highlighted here.

9.4.2 Discussion

As can be seen from the table and project overviews, a wide range of EU projects address flood risk management and spatial planning, covering spatial levels ranging from local to European. All projects focus on river floods, and some also take coastal floods into account as well, but only one project (FLOWS) looked at floods due to heavy rainfall as well.

As expected, most projects focus on the potential of spatial planning to *prevent* floods or *mitigate* flood risks. Even though some projects also go into other components of the DMC, spatial planning is usually not a factor there. Most of the action taken by the projects to increase *preparedness* focus on raising awareness for the issue, but research on the explicit use of spatial planning to improve emergency evacuation is lacking, except for the FLOODsite project, which does look into evacuation models.

Various spatial planning related activities have been (or are being) undertaken by the projects. Many projects included mapping and/or modelling activities, providing cross-sectoral information that is crucial in an integrated SP/FRM approach. Another important aim of many projects is the assessment of spatial *prevention* and *mitigation* measures in order to show the effectiveness of such measures and support the decision making process. The current status of flood management was addressed in two projects. A couple of projects specifically tried to bring together the different actors in flood management in order to facilitate the integration itself and in one project spatial *prevention* measures are actively carried out. Besides the above, many activities not directly related to spatial planning were usually addressed as well.

While in general most projects are research oriented (ESPON, EU-JRC, IRMA-SPONGE, FLOODSITE) quite some projects look towards the real world as well. Some projects looked at current practices (ARMONIA, FLOWS), actively tried to bring together water managers and spatial planners and put integrated flood risk management on the agenda (ELLA, MITCH, ESPACE), or even combined research activities with investments in actual flood management measures (SDF). It is clear that for many EU projects it is important to find a way to put scientific knowledge into practice.

The project activities that have finished all emphasise the need to move from a reactive FRM approach to more pro-active management of flood risks. The value of spatial planning is in this regard acknowledged by all the different projects. The IRMA-SPONGE project showed that upstream flood attenuation measures (i.e. retention areas) mainly have effect just downstream. Furthermore, more can be achieved by *mitigation* then by *prevention* measures with respect to flood risks (Hooijer *et al.* 2004). Clear planning guidelines seem to be necessary, incorporating future uncertainties like climate change, for this to work (FLOWS 2006). The surveys on current practices show that boundary conditions (legislative framework, nature of hazard, etc.) differ a lot between countries. At the moment, flood risk assessments, and consequently management strategies, still rely mainly on hazard information. In cases where spatial planning measures are considered, the respective strategy is usually developed at a higher administrative level (national/regional) then where it is to be implemented (local level). Because of different priorities at these lower administrative levels, many initiatives fail on behalf of shorter-term issues, like unemployment and economic development, as binding legislation is lacking (ARMONIA 2005b). This is implicitly recognised by many of the projects, as many emphasise the need to raise awareness of both public and government (e.g. ESPACE 2006; FLOWS 2006) in order to create social and political basis for future measurements. Additionally, the need for a coordinated multi-stakeholder approach and an open discussion when implementing measures is also a recurring recommendation (e.g. ARMONIA 2005b; FLOWS 2006; MITCH 2007).

Even though a lot of research has been (and is being) conducted, a couple of caveats did emerge while analysing the projects. For instance, even though the climate debate seems ubiquitously, only two projects take climate change into account (ESPACE, FLOWS) and most assessments are done on the current situation, excluding not only climate change but also future economic developments, a critical component of risk. Furthermore, because spatial measures affect large areas, the public is an important stakeholder as well. However, any social consequences of measures, like changes in value of real estate after publication of flood zones, have not been investigated thoroughly. Lastly, most measures that are dealt with aim to reduce direct damage by either preventing a flood or by reducing exposure. However, reducing vulnerability as a means to reduce the impact of a flood (i.e. by mitigating costs with insurances) seems underdeveloped (except in FLOODsite).

9.5 CONCLUSIONS AND RECOMMENDATIONS

This paper looked at the position of spatial planning in flood management and specifically at European Research on this subject. Flood management in Europe is clearly in the transition to a risk management regime where the added value of an integrated approach (IWRM) is widely recognised. The issue is subject to a considerable amount of research. This paper shows that EU projects are involved in both researching possibilities of, and generating information for, an integrated flood management strategy as well as actively facilitating the integration of the spatial planning and flood management sector. The surveys on current practices show that flood risk management is by no means integrated yet, but the various assessments of spatial measures show that a lot can be gained with respect to flood damage prevention when the right spatial measures are taken. Cooperation between spatial planning and flood management is a real possibility, as is shown by various projects, and already integrated measures are taken.

It is encouraging that a wide range of EU activities deal with the interplay of spatial planning and flood management with respect to modern flood management and its solutions. However, this review also brings some shortcomings to light that require additional attention:

- First, currently mainly river floods are addressed in European research, and rain floods are rarely dealt with. Sound spatial planning might be equally promising in reducing the impact of these floods as well, but is hardly addressed yet in research and practice.
- Second, the use of spatial planning with respect to *prevention* and *mitigation* is well studied. However, spatial planning can also play an important role in *preparedness* activities, for instance by facilitating an infrastructural network that can accommodate a large scale, fast evacuation from high-risk areas.
- Third, the actual implementation of spatial measures by EU projects demonstrates that this is not just theory. However, this mainly concerns *preventive* spatial measures, which are still close to traditional (purely preventive) flood management. The implementing of spatial *mitigation* measures has even more potential with respect to damage prevention and would be the next step.
- Fourth, most projects looked at the current situation and neglected future developments like climate change and economic developments. As they will significantly affect risks in the near future, both ought to be addressed in integrated flood management studies.
- Fifth, the surveys on current practices show that there is considerable variation in boundary conditions (legislative framework, geographical setting, nature of hazard, socio-economic structure, etc.) of management measures between countries. A common methodology, as addressed by various EU projects, might therefore not be ideal. Rather tailor-made strategies for specific cases could prove to be more useful, especially considering the transnational nature of many rivers.

9.6 REFERENCES

ARMONIA (2005a) ARMONIA brochure. Armoniaproject, Rome, Italy. www.eu-medin.org/download/leaflet%20ARMONIA.pdf

ARMONIA (2005b) *Report on the European scenario of technological and scientific standards reached in spatial planning versus natural risk management.* WP1: State-of-the-art of spatial planning Deliverable 1.1. ARMONIA Project

ARMONIA Project (2005a) ARMONIA brochure. Armoniaproject, Rome, Italy. www.eu-medin.org/download/leaflet%20ARMONIA.pdf

ARMONIA Project (2005b) *Report on new methodology for multi-risk assessment and the harmonisation of different natural risk maps.* Deliverable 3.1. ARMONIA Project

ARMONIA Project (2005c) *Report on the European scenario of technological and scientific standards reached in spatial planning versus natural risk management.* WP1: State-of-the-art of spatial planning Deliverable 1.1. ARMONIA Project

Blind, M., Hattermann, F., Timmerman, J., and Sikorska, B. (2005) *European IWRM - Research - Outline analysis of projects' contribution for the purpose of selecting topics for synthesis.* Report No 3 of the NeWater project.

Böhm, H.R., Heiland, P., Dapp, K., and Haupter, B. (2002) *Spatial Planning and Supporting Instruments for Preventive Flood Management.* IRMA-SPONGE Project 5. The Netherlands Centre for River Studies (NCR) Delft.

Bons, C.A., Gijsbers, P.J.A., and Schielen, R. (2002) *Interactive Flood Management and Landscape Planning in River Systems:Development of a Decision Support System and Analysis of Retention Options along the Lower Rhine River.* IRMA-SPONGE Project 4. The Netherlands Centre for River Studies (NCR) Delft.

Burby, R.J. (2005) It's Official: Planning Helps to Reduce Property Losses. *Journal of the American planning association* **71**(8), 18-18.

Burby, R.J., Deyle, R.E., Godschalk, D.R., and Olshansky, R. (2000) Creating Hazard Resilient Communities through Land-Use Planning. *Natural Hazards Review* **1**(2), 99-106. http://www.planning.unc.edu/facstaff/faculty/burby/Hazard_Resilient_Communities.pdf

CEC (2006) Proposal for a Directive of the European Parliament and of the Council on the assessment and management of floods.

ELLA (2003) ELLA Preventive flood management measures by transnational spatial planning. http://www.ella-interreg.org/fileadmin/ella_content_material/Informationsmaterial/Flyer/ELLA_Flyer_engl.pdf

ESPACE (2006) ESPACE News 3. ESPACE Project, Environment Department, Winchester. http://www.espace-project.org/newsletters.htm

European Commission (2000) Directive of the European Parliament and the Council establishing a framework for the Community action in the field of water policy. 2000/60/EC. http://ec.europa.eu/environment/water/water-framework/index_en.html

FLOODsite (2005) Research, technological development and innovation activities. http://www.floodsite.net/html/work_programme.pdf

FLOWS (2006) FLOWS - Summary of Conclusions. www.flows.nu

Forbes Tuners, T. (2004) *Flood risk management in FLOWS partner countries.* FLOWS final report 3A. Lincolnshire County Council United Kingdom.

Godber, A. (2002) Urban Floodplain Land-use - Acceptable Hazard? : A Case Study of Flood Risk Perception on the Gurangunbah Floodplain (Nerang River System), Gold Coast. *Proceedings QUT Centre for Social Change Research Conference: Social Change in the 21st Century.* QUT, Brisbane, Australia. http://eprints.qut.edu.au/archive/00000476/

GWP-TAC (2000) *Integrated Water Resources Management.* TAC Background Paper 4. Global Water Partnership Stockholm, Sweden.

Heun, J.C., Schotanus, T.D., de Groen, M.M., and Werner, M. (2002) *STORM-Rhine - Simulation Tool for River Management.* IRMA-SPONGE Project 13. The Netherlands Centre for River Studies (NCR) Delft.

Hooijer, A., Klijn, F., Pedroli, G.B.M., and Van Os, A.G. (2004) Towards sustainable flood risk management in the Rhine and Meuse river basins: Synopsis of the findings of IRMA-SPONGE. *River Research and Applications* **20**(3), 343-357. ISI:000221838800010

IPCC (2001) *Climate Change 2001: The Scientific Basis - Contribution of Working Group I to the Third Assessment Report of the Intergovernmental Panel on Climate Change.* Cambridge University Press, New York.

Kron, W. (2002) Keynote lecture: Flood risk = hazard x exposure x vulnerability. Pages 82-97 in: Wu et al. (ed.) *Flood Defence '2002.* Science Press, New York. ISBN: 1-880132-54-0.

Kron, W. and Thumerer, T. (2002) Water-related disasters: Loss trends and possible countermeasures from a (re-)insurers point of view. Munich Reinsurance Company, Germany.

Lavalle, C., Barredo, J.I., de Roo, A., Niemeyer, S., Miguel-Ayanz, J.S., Hiederer, R., Genovese, E., and Camia, A. (2005) *Towards an European integrated map of risk from weather driven events.* EUR 22116 EN. European Commission - Joint Research Centre

Milly, P.C.D., Wetherald, R.T., Dunne, K.A., and Delworth, T.L. (2002) Increasing risk of great floods in a changing climate. *Nature* **415**(6871), 514-517. http://dx.doi.org/10.1038/415514a

MITCH (2007) MITCH Website. http://www.hrwallingford.co.uk/Mitch/default.htm

Munich RE (2005) *Weather catastrophes and climate change.* Münchener Rückversicherungs-Gesellschaft, München.

Pahl-Wostl, C., Downing, T., Kabat, P., Magnuszewski, J., Meigh, J., Schüter, M., Sendzimir, J., and Werners, S. (2006) *Transition To Adaptive Water Management: The NeWater project.* NeWater Working Paper 1.

Petrow, T., Thieken, A.H., Kreibich, H., Bahlburg, C.H., and Merz, B. (2006) Improvements on flood alleviation in Germany: Lessons learned from the Elbe flood in August 2002. *Environmental Management* **38**(5), 717-732. http://dx.doi.org/10.1007/s00267-005-6291-4

Samuels, P. and Gouldby, B. (2005) *Language of Risk - project definitions.* T32-04-01. FLOODsite Project

Schmidt-Thomé, P., Kallio, H., Jarva, J., Tarvainen, T., Breiving, S., Fleischhauer, M., Peltonen, L., Kumpulainen, S., Olfert, A., Bärring, L., Persson, G., Relvao, A.M., and Batista, M.J. (2006) *The Spatial Effects and Management of Natural and Technological Hazards in Europe.* ESPON 1.3.1.

SDF (2005) Sustainable Development of Floodplains. Rijkswaterstaat-Oost Nederland, Arnhem. www.sdfproject.nl

Siegrist, M. and Gutscher, H. (2006) Flooding risks: A comparison of lay people's perceptions and expert's assessments in Switzerland. *Risk Analysis* **26**(4), 971-979. ISI:000239897400010

Tunstall, S.M., Johnson, C.L., and Pennning Rowsell, E.C. (2004) Flood Hazard Management in England and Wales: From Land Drainage to Flood Risk Management. Proceedings of the World Congress on natural Disaster Mitigation, 19-21 February 2004, http://www.fhrc.mdx.ac.uk/resources/publications.html

van Deursen, W.P.A. and Middelkoop, H. (2002) *Development of Flood Management Strategies for the Rhine and Meuse Basins in the Context of Integrated River Management.* IRMA-SPONGE Project 2. The Netherlands Centre for River Studies (NCR) Delft.

Vis, M., Klijn, F., De Bruijn, K.M., and van Buuren, M. (2003) Resilience strategies for flood risk management in the Netherlands. *International Journal of River Basin Management* **1**(1), 33-39. http://www.jrbm.net/pages/archives/JRBMn1/Rinus.PDF

Vis, M., Klijn, F., and van Buuren, M. (2002) *Living with Floods resilience strategies for flood risk management and multiple land use in the lower Rhine River basin.* IRMA-SPONGE Project 10. The Netherlands Centre for River Studies (NCR) Delft.

Wetterskip Fryslan (2006) *WP1C Best Practise Evaluation.* FLOWS final report WP1C. HydroLogic The Netherlands.

WHO (2002) *Environmental health in emergencies and disasters: a practical guide.* World Health Organisation, Geneva, Switzerland.

Wind, H.G., Nierop, T.M., de Blois, C.J., and de Kok, J.L. (1999) Analysis of flood damages from the 1993 and 1995 Meuse floods. *Water Resources Research* **35**(11), 3459-3465. ISI:000083342400021

APPENDIX 1 – PROJECT INFORMATION

Abbreviation	Full Title	Website
ARMONIA	*Applied multi Risk Mapping Of Natural hazards for Impact Assessment*	http://www.armoniaproject.net/html4/index.php
ELLA	*Elbe-Labe Preventive flood management measures by transnational spatial planning*	http://www.ella-interreg.org/
ESPACE	*European Spatial Planning: Adapting to Climate Events*	http://www.espace-project.org/index.htm
ESPON	*The spatial effects and management of natural and technological hazards in general and in relation to climate change*	http://www.espon.lu/
EU-JRC	*Towards an European integrated map of risk from weather driven events*	http://moland.jrc.it/documents/EUR_22116_2005_Lavalle_et_al.pdf
FLOODsite	*Integrated Flood Risk Analysis and Management Methodologies*	http://www.floodsite.net/
FLOWS	*Floodplain Land use Optimizing Workable Sustainability*	http://www.flows.nu
IRMA-SPONGE	*IRMA-SPONGE Umbrella Program*	http://www.irma-sponge.org/
MITCH	*Mitigation of climate induced natural hazards*	http://www.hrwallingford.co.uk/Mitch/default.htm
SDF	*Sustainable Development of Floodplains*	http://www.sdfproject.nl/

10

The social dimensions of IWRM: Where does inequality fit in?

Rhodante Ahlers and Dagmar Ridder

10.1 INTRODUCTION

A more integrated and adaptive approach to water management generally includes a concern for the poor and marginalised. Often poverty and gender are linked with health so as to capture the more vulnerable sectors of society. These sectors are argued to need special attention to avoid mistakes made in the past and to alleviate the lack of access to clean and sufficient water for basic subsistence and income opportunities in the future. Research on the issues related poverty, gender and health in the water sector have been highly compartmentalised, historically due to the engineering focus of the water sector, and currently due to the same engineers arguing that they lack the necessary expertise (Zwarteveen 2006). Unfortunately the latter point has only minimally led to more research and educational efforts to develop this expertise in the water sector. While the discipline of economics, and its partners business administration and political science, have been embraced and even celebrated since the early nineties, this is less clear for those disciplines that explicitly study the alleviation of inequalities such as sociology, gender studies and political ecology. This chapter intends to raise the relevance of the social dimension in water management. Rather than discuss gender, poverty and health separately, we focus on the dynamics of social inequality as the underlying force behind gender, race, and income inequality that lead to inequitable access to water for economic and health purposes.

This chapter discusses how research funded by the EU on water management and IWRM in particular has integrated social dimension in its endeavours. To do so we have screened 60 research projects under the Directorate General (DG) Research, even though others, such as the 'DG Environment,' also include IWRM related research. First the research projects under the key action water within the Environment and Sustainable Development Programme (5th Framework Programme) were screened for whether social dimensions were relevant to their topic. Secondly, of the 35 remaining projects only those were considered if they dealt with integrated water (resource) management, if they had a social or human dimension of water management as (one of) their objectives, and were not be completed before the year 2000. Thirdly, of the remaining 25 projects we discuss six in detail identifying to what extent their objectives and inclusion of social dimensions translate into an analysis of social inequality. For the review to be manageable, we made the choice to review projects that made a concerted effort to include social dimensions. A full review of those projects that should have considered social dimensions but did not, lies outside the scope of this chapter.

Rather than focus on gender or poverty alleviation or health, we argue for a more integrated approach that analyses the multiple social and economic dimensions of inequality that interact with each other and the biophysical context. Thus we step away from the neutral and often vague terminology of poverty and identify the primary forces of inequality as shaped by gender relations, economic inequality, and the inequality between the global North and South. Hereby we want to make explicit the articulation of different economic, administrative and political scales that affect inequality, as well as the importance of contextual accuracy. Inequality is complex but does not have to be vague. In

 The adaptiveness of IWRM; Analysing European IWRM research Edited by Jos G. Timmerman, Claudia Pahl-Wostl and Jörn Möltgen. ISBN: 9781843391722. Published by IWA Publishing, London, UK.

accordance with the EU review of water research from FP4 to FP6 that argues that IWRM is an inherently political process (Gyawali *et al.* 2006), we argue that when making the politics behind water management decisions explicit the social dimensions become strikingly clear, and inequality in decision-making and allocation can no longer be analysed as merely unintended consequences. In sum, if we are serious about addressing inequality in access and control over water, the underlying mechanism of these inequalities should be analysed and tackled.

After providing a contemporary analysis of these forces of inequality within the water sector, we proceed to review the selected EU funded research projects. We discuss 6 projects in detail in order to identify the new insights that have been developed and the necessary gaps that still demand urgent attention. We conclude with a discussion arguing that lip service to equity will not only reproduce old forms of inequality but may create new ones. An integrated approach needs to take inequality seriously and invest in both educational and research development that focus on inequality as a central concern of water management. This should not only be of importance in the global south but equally so in the global north.

10.2 INTEGRATING THE SOCIAL DIMENSION IN WATER MANAGEMENT

A more integrated approach to water management has come with a demand for more inclusive and participatory water governance, a recognition of the complex nature of socio-ecological systems and increasing uncertainty requiring more adaptive management and an inclusion of diverse skills and knowledge sources (van der Zaag 2005 and see Chapter 2). Despite the understanding that the social dimensions in water management are self-evident and crucial, very little structural effort has been made (or funded) to develop this understanding. Of the social dimensions gender has gained increasing currency. In the recent past, gender analysis of water management policy has resulted in conceptual insights for improving both technical and institutional aspects of water resources management in both the agricultural and municipal sector. With regard to poverty alleviation much focus has been on the MDGs and pro-poor policy. Unfortunately, within IWRM this synergy is only just gathering momentum. Very little conceptual research is available in the water sector on the contemporary dynamics resulting from the politics of the Global North in relation to the Global south[1]. In the water sector only the privatisation endeavour in the drinking water sector has referred to this dimension of inequality (Bond 2004; Swyngedouw 2005).

Identified as the principle lesson from the EU INCO water research FP4-FP6 Review (2006) is that IWRM is inherently a political process as decisions, policies and practices are led by insights coming from different epistemic communities and priorities defended by those with decision-making power (Gyawali *et al.* 2006). Hence, for effective IWRM informed by stakeholder participation an understanding of these social political processes is crucial. It is this integration of the social with the technical that has proven to be a challenge: "Statements about water users are often about how engineers and planners hope or wish them to behave rather than about observations and understanding of actual practices and experiences" (Uphoff (1996) cited in Zwarteveen 2006:31).

Gender and economic inequalities have generally been dealt with as points of attention, with scant analyses of their underlying causes or their dynamic link and therefore structural impact on water management. If we agree that IWRM is also very much a political process, then its social dimensions need to be analysed as equally political. Hence, gender and income, and global inequality no longer remain as characteristics of a given context but as drivers for the political process that shapes IWRM.

Even though the concept of gender is now well known and widely recognised as a salient dimension in water management, its original content has been compromised. While the term is globally accepted and to a certain extent espoused by different disciplines, professions and sectors at all institutional levels, the change in gendered access to resources falls far short of the expenditure invested in it. In gender mainstreaming, the concept of gender deals mainly with the different roles, and subsequently needs, that women and men may have. Gender is often misunderstood as a focus on women, or as a focus of women and men being very different. Stereotypical and gendered notions of productive and domestic water use has in the past lead to a focus on "women and water and sanitation" alongside "men and irrigated agriculture", often replicated in decentralised participatory approaches. Not only does this reproduce disempowering gender stereotypes, ignoring the knowledge and labour women contribute to agricultural water management, it also fails to recognise uses of water that may be essential to livelihood securities. Equating gender with women disconnects women and men from the social context in which gender relations are shaped and reproduced with the danger of legitimising stereotypical and a-historic notions of men and women's roles (Ahlers 2005).

Instead, gender concerns the complex everyday societal organisation of roles and responsibilities between men and women, shaped by the political, economic and cultural dimensions of social relations that have both particular, as well

[1] Analyses in this sector is far more developed on governance in general (structural adjustment and conditionality of lending programs) (Arrighi 2002; Arrighi and Silver 1999; Escobar 1995; Hoogvelt 2001) and resources such as oil (see the work by Zalik (2004) and Watts (2004)).

as structural expressions of cooperation and conflict. In this sense women are not a fixed category with needs that always and everywhere differ from those of men. Rather they form a heterogeneous group that shares certain common interests due to processes of structural inequality in society. Economic and gender inequality should therefore be understood as interdependent, and intersecting with social disparities shaped by notions of class, race, caste, ethnicity, age, ability, and religion. These processes are highly influenced by income and food (in) security. The Gender and Water Alliance (GWA) has compiled a resource book on gender and IWRM in which the core principles of IWRM are discussed to show how gender inequality may affect women's access to water and decision-making processes over its management. GWA argues that at each step of the process, the implications for both men and women should be assessed so that men and women's experience form an integral dimension of the design. Not only is this to improve the effectiveness and efficiency of water sector programs, it is argued to contribute to more gender equality (GenderandWAterAlliance 2006).

Poverty is a complex phenomenon but is primarily an outcome of unequal access to resources. The joint agency paper (including the EU) on Poverty Reduction and Environmental Security (UNDP 2002) identified three key dimensions to poverty reduction: livelihoods, health and vulnerability. This framework is argued to provide a basis for the analysis of the relationship between water management and poverty reduction, with different actions in water management addressing different dimensions of poverty.

Both these approaches allow for an incorporation of gender and poverty concerns in IWRM but primarily focus on making the issues visible or in their words "assessable." It remains less clear who will do the assessing and to what extent the assessment carries sufficient authority (besides funding conditionality) to affect change. A further limitation with these frameworks is that women and men, poor and rich are identified as clear categories. While this may be politically useful it is unfortunately a simplification, which does not do justice to the complexity and contextuality of inequality. Therefore, we suggest a focus on the social material practices that allow gender and income inequality to change or persist, as well as be reproduced within contemporary policy reforms. We do this by focussing on three specific issues that are central to IWRM: water re-allocation, water pricing and participation.

Water reallocation

Given increasing competition for water and agriculture being the larger user of water, the prospect for intersectoral reallocation of water has surfaced as a possible measure to alleviate thirsty cities and increase the productivity of water in an economic sense, both in terms of increasing water productivity in agriculture as well as water productivity in general. A recent comprehensive study by IWMI (Molle and Berkoff, 2006) shows that it may not be water cities are lacking but the economic means to deliver it. Neither is there strong evidence that the other competing sector, industry, lacks water to continue or even to expand their activities. Hence the large economic gains anticipated by inter-sectoral reallocation appear to be overstated, especially when taking into consideration high transaction costs in a context of unclear or politically sensitive definitions of water rights (Perry 2001; Savenije and Zaag van der 2002; Young 1996). The environmental and social costs however may be detrimental in the long run due to several factors generally underestimated:

(1) The importance of rural agricultural employment for income security and mitigation of rural out migration
(2) The importance of low value staple crops for food security especially in areas of high HIV AIDS incidence
(3) The multiple uses and users of water not recognised by the formal economy or by formal water management institutions,
(4) Legal pluralism of water rights and in particular the lack of recognition of informal or indirect use rights, and common property arrangements.

Transferring water out of agriculture without due attention to the economic impact of the communities surviving on irrigated agriculture not only economically but socially disables these communities. Shifting water to higher value crops may increase economic productivity of water, and to a certain extent agricultural productivity, it does not necessarily ensure food security: high value uses such as recreational use (e.g. golf courses) and horticulture generally do not service the rural or urban poor (Ahlers 2005; Frederiksen 1996; Young 2000).

With regard to water rights, transfers have in the past considered compensation packages. However these do not take into account the legal pluralistic character of water rights in practice. Water rights are generally dynamic and context specific. Furthermore they are often linked to other rights or encompass multiple sources. Hence rights are usually not clear, specific or singular, but carry multiple claims that may have both temporal and spatial dimensions (Meinzen-Dick *et al.* 1997). Certain groups have either been expropriated from their rights to land and water (the indigenous and colonised) or were not identified as water users. Rights to land and water have been acquired through force, by sale or assigned through land reforms. The first two options are not in reach for the poor, while in the third women were generally excluded. Due to historic gender inequities, formal rights have been in the past allocated to heads of households. In land reform programs and irrigation projects these household heads have been invariably defined as male (Agarwal 1994; Ahlers 2002; Deere and Leon 2001; Zwarteveen 1997). Historically therefore, the poor and women have negotiated other routes to access water to secure their production and consumption interests. These routes do not carry the necessary formality to allow them to be compensated, or to be invited to participate in decision-making over reallocation schemes, nor to meaningfully participate in water markets (Ahlers 2005; Zwarteveen 1997).

Reallocating water out of agriculture or away from certain crops may make economic sense and perhaps even lead to higher efficient water use, although little empirical data is available (Molle 2006). However, when taking social inequality in water management into consideration such reallocation demands new scrutiny in relation to food and income security.

Water pricing

Water pricing has proven to be economically complex and socially and politically sensitive for the domestic water supply sector. The focus has been primarily on cost recovery and asset management. For the agricultural sector the purpose is both cost recovery and conservation, as pricing is to induce more prudent use or increase the economic productivity of water. Evidence for both are sorely missing: capturing environmental and social costs are still underdeveloped and rarely consider the extractive and in situ values of the resource base itself (Turner *et al.* 2004), while the economic costs due to resulting political instability are often ignored even though these may be tremendous as the Bolivian water war cases of both Cochabamba and La Paz demonstrate so clearly (Ahlers 2005; Kohl and Farthing 2006; Laurie 1999; Spronk 2006).

At the centre of the complexity of water pricing lies the complexity of water itself: its fluidity, its variability, its public good and common pool characteristics, and finally: water flow reacts slowly to alterations. The gender and poverty dimensions of each of these are partially known and are highly context specific dependent on hydrological parameters as well as on historic management institutions guiding allocation and distribution. Strict Operation and Maintenance (O&M) recovery rates that include capital depreciation and replacement costs is prohibitive for competitive farmers in the OECD countries let alone marginalised farming households elsewhere (OECD 2002). Where water is less reliable or more difficult to access (either due to physical or political hindrance) O&M costs will increase (infrastructural, energy, labour, etc.) while at the same time it is under these circumstances that water security is a priority. This, however, does not imply that water should be supplied for free. What it does mean is that safeguarding accessibility to high quality water for basic needs enables food and income security for the poor and prevents public health costs related to water borne diseases, as well as diseases resulting from a lack of water. The case of the cholera outbreak in KwaZulu Natal in 2000 after several thousand families had been cut off due to non payment is a case in point (Mckinley 2005).

From a gender and poverty perspective the debate between willingness and ability to pay is often raised. In this respect gendered access to the labour market and financial sources such as credits in combination with the gendered labour division in domestic water related tasks translates into gendered discrimination in access to water. Poor women are often very willing to pay for dependable services, certainly if this means less time and less humiliation in negotiating a reliable, timely, and clean supply of water. However, this does not mean that they can sustain the payment of higher water prices over the long run. Given the increasing importance of the informal economy with rising precariousness of income, access to health and food security may be jeopardised if water bills increase at the same time. This is compounded by the fact the poor and poor women in particular, historically have less access to property rights, financial and political institutions, in addition to constrained access to decision-making bodies and sources of income and information (Ahlers 2000; Cleaver and Elson 1995; Zwarteveen 1997). In agriculture ability to pay is also largely dependent on factors outside the control of small-scale producers such as world market prices, or subsidy and tariffs politics.

Consequently, focussing on safeguarding access to good quality water for all its uses, rather than increasing its rather narrow economic return, may be more appropriate, especially given the MDG of increasing and securing access to clean water. Engendering competition between the different water sectors, or the private and public sector distracts from achieving better access for all, and carries the danger of overly fragmenting the water sector rendering it less rather than more efficient (Bakker 2003; Young 1996). Especially if economic return does not reflect the informal sector, or the potential health costs resulting from lack of water, or the use of second or third grade water. Besides the health effects on the producers and consumers, the economic effect of destroying market potential can ruin the livelihoods all along the commodity chain. Often neglected is that women who have to deal with contaminated water also expose their unborn children and those they breastfeed. Finally, with regard to considering environmental needs (or flows), the field of political ecology emphasises the notion of nature as a result of by bio-physical and human interaction. Hence, understanding the water scope as a socio-ecological transformation lies much closer to an integrated notion of water management.

In areas of high HIV and Aids incidence the issue is even more pressing and more complicated because available adult labour power is limited, and sufficient nutrition is crucial. Providing these stressed households with sufficient water should not be limited by fixed household quantities (albeit at a reduced price) or volumetric pricing that is prohibitive for basic needs and securing the nutritional levels so direly needed.

Contemporary water pricing needs to link partial cost recovery with social and economic responsibility along with environmental needs, in the sense that water, food, and health security for the poor allow for stable economic activity in a healthy environmental context. Water pricing will only work if it is flexible, context specific and highly sensitive to historic inequities and contemporary social dynamics.

Participation

Institutional reform in the water sector places much emphasis on participation and multiple stakeholder platforms. Often though these formal institutions are foreign to the context in which they are to function. Furthermore, decision-making procedures have social and political dimensions that are dynamic and context specific and generally harbour power differences according to economic, gender and ethnic/race/caste differences (Cleaver 1998, 2000; Jackson 1998; Mayoux 1995). With respect to water, access to decision-making bodies usually embodies formal recognition as a water user while historic inequities have hindered certain groups in society to such access (Ahlers 2000; Zwarteveen and Neupane 1996). While the international consensus argues for increased participation of previously excluded groups with much effort geared to encourage these groups to invest their time and energy in different decision-making bodies, we are slowly learning that a seat at the table may not at all mean influence in the decisions made due to power inequities between stakeholders, often induced by naïve notions of homogenous and harmonious communities (Ahlers 2002; Baxter and Richard 1990; Cleaver 1999; Mamdani 2000). Also, the particular table to which one is invited may not be the one at which final decisions are made.

A more participatory and gender balanced management of water resources is now a quite established policy directive supported by (I)NGO's and established water policy and financing institutes. In practice this has been translated into making women and their uses of water visible in the water sector and by encouraging women to participate in decision-making bodies related to water resources management. Increased involvement of women in male dominated institutions was perceived to address gender inequality in access to and control over water resources. Unfortunately experiences over the last ten years are quite ambiguous. Increased participation of women in formal water resource institutions does not show a very encouraging trend. Studies from Nepal, India, Zimbabwe, Mexico and Bolivia show little evidence that participation is per se empowering or positively impacting water management (Meinzen-Dick and Zwarteveen 1998; Toner and Cleaver 2006; Zwarteveen and Neupane 1996)

Much has also been achieved: although primarily at the bi- and multilateral level women are now recognised water users, there is a seat now available for them at the table, and this process is financially supported at most institutional levels. This is not so much the case for the poor. Gender advocacy in this sense has been far more successful. Nevertheless, the question arises however, if 'the seat at the table' actually translates into control sharing (Cleaver 2003; Cornwall 2003). Even if the decision-making bodies may have become more gender balanced, the power to influence decisions has not. Without the enabling circumstances that make participation meaningful, having a seat at the table may translate into a waste of time. The emphasis should be on how the newcomer sits at the table instead of how many. Time spent in decision-making procedures where they are not heard, means less time for the activities formerly engaged in to secure access to resources. Finally, the table may not be appropriate for addressing the social dimensions such as gender and poverty. We may need to recuperate or develop new institutional forms that create space for men and women to participate without reproducing or legitimising old stereotypes (Cleaver 2003). For these institutions to function they demand legitimation only possible if close attention is paid to effective representation and appropriate decision making procedures.

The possibility to participate, to be heard, to be listened to, and to influence decision-making has historically been strongly gendered in the global north as well as the global south. Membership of the 'waterschappen' in The Netherlands for example was originally based on property ownership. Property ownership in the Netherlands due to its protestant Christian influence was highly skewed towards men. Encouraging participation without acknowledging power differentials will only lead to reproducing and legitimising inequality (Mearns *et al.* 2000; Wester and Warner 2002).

In sum, participation as a dimension of IWRM is a widely advocated but far from self-evident process, particularly not for the poor, and for those able to participate it may bring high social and economic costs (Mayoux 1995). Of importance is

- The recognition of difference by identifying conflicting and common interests in a context riddled with power differentials;
- The recognition of potential costs for different stakeholders, and how benefits may be highly differentiated;
- Linking participation with structural change: incorporating underlying inequalities in access to resources and decision-making power.

Research is needed on who can and does participate in designing IWRM strategies, to what extent this is inclusive of all social categories, or is limited to only certain sectors of society, or defined by certain geographical location, primarily the global North. If we take the example of the water professionals: the majority are still male, and hardly reflect the diverse cultural make up that Europe has become. This is not unique for the water sector, but an issue European society is increasingly been made aware of.

Crafting institutions appropriate for IWRM also demands sensitivity to the changing demands from society, and changing citizen rights and responsibilities (Barham 2001). Citizenship is currently in flux and alongside those in society recognised and provided with full rights, a significant number of people move through Europe without these rights, often invisible but very much in need of sufficient clean water. Even though we focus on water supply and sanitation challenges in slum areas in the global south, we give little attention to similar developments in the global north.

Hence, if participation is to play a central role in IWRM, we need to understand and be willing to change the existing rights, roles and rules in so far that they allow social and material practices that reproduce inequality. By extension, this means a critical review of the politics of policy formulation on a global scale, seeking a change in the dominance of the global north in policy identification on the one hand and recognition through research and development funding on the other.

10.3 ANALYSIS OF SOCIAL INEQUALITY IN SELECTED EUROPEAN UNION FUNDED WATER MANAGEMENT PROJECTS.

Understanding the social dimensions of water management has increasingly gained recognition as being essential in effective water management, and has strongly informed the millennium development goals. Unfortunately, until recently, we hardly see this translated in European funded Water Management Projects. Where funding is destined outside Europe or geared toward Southern Mediterranean or Eastern Europe, social issues do appear on the project agenda. For projects elsewhere in Europe, however, these issues are rarely considered.

How ever, this does not mean there is no poverty in the north that participation is not gendered, and health issues are irrelevant. We argue that these social dimensions are relevant and essential for water management everywhere.

10.3.1 Selection of projects to be analysed

Initially a selection was made complying with the following three minimum criteria:

(1) They should deal with integrated water (resource) management,
(2) They should consider the social dimensions of water management and
(3) They should not be completed before the year 2000.

On the basis of these criteria the selected projects belong to the EU 5th RTD framework programme "Energy, Environment and Sustainable Development (1998-2002) and the 6th RTD framework programme "Sustainable Development, Global Change and Ecosystems" (2002-2006). Water-related projects funded by the EU within the 5th framework programme did not specifically recognise the need to address the issue of poverty, gender or health. Accordingly, information and data is scarce and make a coherent comparative analysis of projects impossible. Even EU projects funded within the 6th framework programme, which included the obligation for each project to define its gender action plan, lack a coherent strategy to implement it. Addressing gender is generally translated in identifying women's needs and limited to selected subtopics, thereby legitimising gender stereotypes. It is also highly dependent on the motivation of individual researchers to implement a gender approach.

Therefore, a more refined selection of projects out of the 3 above-mentioned programmes followed, with using the subsequent criteria:

- Specifically addressing poverty **or** gender **or** health **or** a combination;
- Indirectly addressing poverty and gender by dealing with e.g. participation;
- A gendered impact is to be expected;
- The impact is expected to be socially differentiated;
- Good accessibility to data/information for the authors.

This process resulted in the following projects:

(1) The FP5 programme included a cluster of projects on the social dimension of water management. Out of this cluster we chose **HarmoniCOP, Euromarket** and **Intermediaries**;
(2) Within FP6 we selected **NeWater** and **Aquastress**;
(3) A project with a specific focus on gender was **Gewamed** (Mainstreaming gender dimensions into water resources development and management in the Mediterranean region).

10.3.2 Analysis of projects regarding the integration of social dimensions.

The following criteria for analysis were applied: Do projects acknowledge the importance of specifically addressing social dimensions? To what extent do they specify their approach?

The analysis differentiates between the intensity of recognising social inequality and integrating issues such as income and gender inequality into the project design. We used a scale from 1 to 3 where:

1 = Project description and/or project results mention the need to consider issues of equity or gender in particular;

2 = Projects include a selected theme/group/work package dealing with equity or gender in particular;

3 = Projects include a coherent gender approach for the project, and equity issues are considered in an integrated manner.

Having identified the contemporary issues in relation to social inequality in Section 2, the projects were also screened in how they deal with issues such as water pricing, water reallocation and participation. The table below gives an overview of these projects and how they have been classified based on their integration of social dimensions. It also shows to what extent they have recognised the issues discussed in Section 2.

Table 1: Overview of the projects equity with classification in integration equity and/or gender in particular

Project Name and Duration	Full Project Title	Research Theme	Recognition of water re-allocation, water pricing or participation?	Classification
HarmoniCOP 10/02 – 12/05	Harmonising Collaborative Planning	Increasing the understanding of participatory river basin management in Europe. Focus lies on conceptualising social learning and active participation in context with the WFD.	The project focus is clearly on participation. Water pricing is considered during conceptualisation but water re-allocation is out of scope.	1
Euromarket 01/03 – 12/05	Water liberalisation scenarios: An empirical analysis of the evolution of the European water supply and sanitation sectors	Analysis of the consequences of liberalisation of water supply and sanitation sectors in economic, ecological, social and institutional terms by means of exploring different liberalisation scenarios.	Water pricing is central as is the role of stakeholders. Water re-allocation was not directly mentioned except for a brief discussion on the elasticity of demand.	2
Intermediaries 11/02 – 10/05	New intermediary services and transformation of urban water supply and wastewater disposal systems in Europe	Assessment of the contribution of "intermediary" services, emerging between the user and provider of water supply and wastewater disposal facilities, to the more rapid implementation of key objectives of EU water policy	Water pricing and the role of stakeholders in analysing new organisations and institutions. Water re-allocation could have been discussed but was not.	1
Aquastress 02/05 – 01/09	Mitigation of Water Stress through new Approaches to Integrating Management, Technical, Economic and Institutional Instruments	The Integrated Project will deliver enhanced interdisciplinary methodologies enabling actors at different levels of involvement and at different stages of the planning process to mitigate water stress problems.	Water allocation and water pricing are investigated under the heading "water saving". Participation is analysed and its practical application a requirement in the case studies.	2-3

Table 1 (continued): Overview of the projects equity with classification in integration equity and/or gender in particular

Project Name and Duration	Full Project Title	Research Theme	Recognition of water re-allocation, water pricing or participation?	Classification
Gewamed 15/02/06 – 14/02/10	Mainstreaming gender dimensions into water resources development and management in the Mediterranean region	To contribute to the mainstreaming of gender dimensions in the Integrated Water Resources Management (IWRM) in the Mediterranean Region by creating dialogue, establishing networks, building a knowledge base and by establishing the structures for promoting the adoption of gender sensitive policies.	Participation is crucial for conceptualising and implementing gender mainstreaming and therefore part of the projects' objectives. Water pricing and water allocation will be rather covered through collaboration with other projects but due to the recent start of the project this is hard to judge.	3

HarmoniCOP (www.harmonicop.info)

The main objective of the HarmoniCOP project has been to increase the understanding of participatory river basin management in Europe and to generate practically useful information on, and improve the scientific base of, both social learning and the role of IC tools in river basin management (see also Chapter 1). The focus of HarmoniCOP on social learning by active participation includes the stimulation of collective action and the resolution of conflicts. This requires the stakeholders of a participatory process eventually aiming for joint action to recognise their interdependence as well as their differences and learn to deal with them constructively. The different groups are to learn and increase their awareness about their biophysical environment and about the complexity of social interactions. Characteristic of the differences identified concern gender and income inequality.

The focus of HarmoniCOP was to help with the implementation of article 14 of the European Water Framework Directive. Hence, the projects' main target groups were accordingly practitioners in water management from the regional to the local level. Still, during the conceptualisation of social learning, the diversity of the participants became a major topic of discussion due to importance given to integrating all groups in their diversity into the participation process. This included specifically economically vulnerable groups including women and men, but also the elderly. The HarmoniCOP handbook "Learning Together To Manage Together – Improving Participation In Water Management" mentions that during the selection of potential stakeholders for the planning process also those who do not have the necessary resources to participate need to be considered (Ridder *et al.* 2005). Time, money and being away from home may be constraining factors for participating in the process. Another dimension refers to resources in terms of information and knowledge. It is argued that groups with less financial means have less access to information sources and knowledge development.

The conceptualisation of social learning (Craps 2003) puts much emphasis on the analysis of the social context in which water management takes place, also as a necessary input for stimulating effective and active participation. In this participation phase differences in gender and social status are explored in regard to recent water management practices and potential change. The project explicitly incorporates social dimensions, and recognises them in terms of inequality. But, rather than making inequality a focus of the project, it is discussed as a methodological challenge. The HarmoniCOP handbook as the most important project output for practitioners, could have included 1) an analysis of inequality so as to identify what the social and economic costs are for different participants and 2) that participation is riddled with power differences and hence involves common and conflicting interests. These are not only solved by designing better participatory mechanisms, but demand linking participation with meaningful action to overcome this inequality. Only then will promoting active participation and social learning become meaningful for an improved water management in Europe.

Euromarket (www2.epfl.ch/mir/page18246.html)

The EUROMARKET project began in January 2003 and focuses on Water Supply and Sanitation services with a distribution network for urban and rural uses of water. The objective of the project is to provide insights concerning the possible liberalisation of water services in the European Union. Besides looking at European policies alongside national organisations and markets that influence water services, the project also assesses the potential impacts of different liberalisation scenarios. These include possible social impacts. Eventually the project will provide recommendations for developing a European policy for water services in Europe. The project endeavours to fill a theoretical gap in water literature on understanding the interactions between the water supply and sanitation sector's dynamics, the enterprises behaviours and strategies, and the emerging regulatory services (Euromarket 2003b).

The project results show that there is no clear indication that European policies are successful in influencing the Water Supply and Sanitation (WSS) services towards liberalisation. Instead drivers and constraints can be identified that encourage a more liberal and competitive WSS sector in Europe. Three major topics were identified that may have a strong impact on liberalisation (Euromarket 2003a):

- The emergence of new stakeholders, with on the one hand the development of (public and private) commercially oriented companies in the water sector, (whose objectives are to widen their area of activity), and on the other hand, the suspicion of public opinion about the profit maximising interests of these generally monopolistic companies;
- National, European and international discussions on the nature of WSS services;
- The evolution of the political and juridical framework that has allowed the development of public-private partnerships.

Here, we focus on the social dimension in relation to the liberalisation processes. The project puts equity as a concern in its work description and deliverable no. 8 addresses this problem specifically. Questions such as "...should everyone have access to the same service?" or "...what is a fair distribution of services?" direct the analysis of the impact of liberalisation scenarios on society.

The project shows that Public Participation can be a driver of liberalisation with a public that supports it (e.g. because people see the possibility of decreasing prices) or with a public that opposes it (e.g. because they fear a negative impact on the environment) (Euromarket 2003a).

The project points out the importance of the provision of clean water to everyone and the importance of sewage disposal as an enormous positive externality to public health through the control of water-borne diseases (Euromarket 2003a). Unfortunately the project did not follow up in detail how different liberalisation scenarios may have a different impact on public health because of differences in access to WSS for various social strata of the public.

For the analysis of equity two main criteria were chosen: *employment* and *access* (Euromarket 2005). *Access* includes rate of connection, social funds, social policies of companies, even (or uneven) distribution across different geographical areas, and tariff principles. *Employment* includes the number of workers employed, the nature of employment contracts, working conditions, and equal rights of workers. With the means of these social indicators the two criteria were applied for different European countries.

A problem inherent to an evaluation based on scenarios and making use of case studies is on the one hand the rather abstract level at which it takes place, and on the other the difficulty to compare between cases. For example, the acknowledgement of the existence of social funds and tariff principles that consider equity aspects does not say anything about their effectiveness on the ground under different conditions. Here, only in-depth studies can be applied that scan the conditions under which the policies in question work. It represents a general problem of many comparative studies of European countries that rely on highly aggregated indicators that directly relate to the policy level of the respective countries. But often these indicators do not show the reality of the complexity policies work on the ground due to their limited, biased or lack of implementation, or to unintended outcomes.

The analysis depicts that different liberalisation scenarios do not impact on the distribution of work among women and men – women are generally poorly represented in the water sector whatever the management mode be. This looks different when analysing the implications of social tariff principles and social funds within different scenarios. The project argues that scenarios representing a high degree of liberalisation are not always disadvantageous for poorer households. For example, outsourcing to more cost effective contractors may lower tariffs and therefore improve the affordability of water services for poorer households. This obviously assumes that cutting costs is translated into lower water tariffs, which is not necessarily what happens in practice. They also point out that community management of water services may not always benefit poorer neighbourhoods as the community with more human and financial resources will be more capable in investing in organisation and management.

Projects like Euromarket depict how social issues could be incorporated into water research if defined as important. But the study suffers from a lack of conceptual insight in the issues related to income and gender inequality (for example the undifferentiated community) but also demonstrates the difficulties of working with gender-related indicators because gender disaggregated data provision is still very limited – regardless if the water service is provided by the public or private sector.

The concerns related to water pricing as discussed in Section 2 are given some consideration in the project but the analysis is devoid of the interrelatedness of social and environmental needs in relation to the cost recovery principle and perceives community as homogeneous. In certain urban settings where affluent neighbourhoods are more clearly demarcated from poor neighbourhoods this may be the case. It remains therefore a public responsibility that poor neighbourhoods are equally serviced. A serious concern is the assumption that increased competition translates into increased efficiency and is therefore beneficial for the consumer as it ought to result in lower tariffs. Both assumptions have not been empirically proven in practice as of yet.

Intermediaries (www.irs-net.de/intermediaries)

The Intermediaries project, starting in November 2002, investigates the contribution of intermediary organisations to the sustainable management of urban water resources in Europe. The project maps the development of intermediary services and organisations in water supply and sanitation services, examines how they facilitate the application of new resource-saving technologies and social practices and assesses their impact on the environment, economic efficiency and infrastructure management (Moss *et al.* 2005). The intermediaries identified are very diverse: from commercial companies or non-for profit organisations to service providers and distributors to educational organisations and consumer group. Despite their differences they have one thing in common: they work between the utilities, regulators and consumers of water services and carry out activities somehow related to water management. The working definition of intermediaries in the project proposal is: "intermediaries are institutions that through various novel styles of policy action introduce new technologies and changed social practices into the production-consumption relationship to reshape the intensity, timing and level of water use and wastewater production" (Moss *et al.* 2005:4).

Social inequality in communities, neighbourhoods and households were not specifically investigated within the scope of this project. To certain extent income inequality played a role in the analysis as a social driver for the emergence or development of intermediary organisations. As intermediaries are also active in the field of public participation they have the potential to – and some intermediaries do - influence different social strata to participate in water management decisions, disseminate information on water saving technology for example, or raise awareness on water-related problems (Medd *et al.* 2003). But the poorer social strata were informed rather than forces of change as the impetus to act came from the outside. How the poorer sectors of society are, or could potential become, drivers of intermediary organisations and therefore influence the water management is far from self-evident and demands further investigation. One problem is the lack of resources of the poor to organise and become actively involved. What is often not recognised is the less formal or alternative organisations of these actors who have lost faith in the mainstream avenues for change. Furthermore, their interests are often not considered as important and thus they do not see the point of investing their scarce resources. As mentioned earlier, the experience of participatory approaches has taught us that the more affluent can not only afford to invest in information but are also more comfortable in attending meetings and speaking up. Participation is not a power free exercise and therefore remains far more attractive to those who see a considerable chance of their interests being attended to. An analysis of intermediaries that better integrates these aspects could benefit the research in social dimensions of water management in Europe. In particular the project question of how intermediaries have an impact in changing social practices and introducing new technologies could be very well followed up with a gender approach that recognises income and other social inequalities. Exploring how social processes of change can be stimulated, it is essential to differentiate among all the actors involved or affected.

NeWater (www.newater.info)

The NeWater project is based on the hypothesis that IWRM cannot be realised unless current water management regimes undergo a transition towards more adaptive water management. Given the complexity of integrating different water demands and the increase in uncertainties in the light of global change NeWater tries to identify a new approach to guarantee sufficient water of satisfactory quality for competing demands and the protection from water related risks. One of the key elements is the consideration of good governance which entails e.g. a fairer distribution of and access to resources.

Within the NeWater project one work package specifically deals with the social dimensions of IWRM: poverty alleviation, gender and health. One result is a new conceptual framework: the LASER framework (Learning from and Adapting to Social-Environmental Realities). "The framework is based on the premise that the aim of an adaptive water management system is to enhance human well-being from adequate access to water and reduce human ill-being resulting from water related hazards. Given the importance of water in alleviating poverty, redressing gender inequalities and improving health conditions, it can be argued that a net gain in human well-being will be achieved by enhancing the capabilities of individuals and communities for water management" (Sullivan *et al.* 2006: pg 13).

The NeWater project therefore explicitly follows an integrated approach and incorporates gender and equity concerns. However, even though equity is defined, along with economic efficiency and environmental sustainability, as one of the three pillars of IWRM (see Fig. 1, in Chapter 1), during project implementation it has received far less attention. Due to the working mode of EU projects, approval of project applications demands that one has to deliver a series of new research results in a comparatively short time. This contradicts working in a participatory manner. Because participatory activities were required at an early stage the later developed LASER framework was almost exclusively applied for research on poverty, health and gender and not as a general framework for participatory activities dealing with research on water management. A second problem encountered was that the planning phase rarely gives sufficient time for stakeholder consultation. As funding is limited over a defined period, projects need to deliver outputs almost immediately. Integrated and very comprehensive projects that work in a participatory manner have to be continuously re-adjusted according to the time schedule of stakeholders. Also the desired stakeholder input may jeopardise project planning. Therefore, the logic of the planned project schedule where much research is produced in parallel rarely corresponds with the logic of implementing participatory projects step by step in the field.

As a result the NeWater project offered research on poverty and gender on a voluntary basis – only when stakeholders considered it important for their case studies. It was not foreseen that a stakeholder awareness raising may be necessary for the project success. This may lead to the situation that social dimensions are neglected and raises important questions about project responsibility in agenda setting and facilitation. Including social dimensions in water management projects and demanding sensitivity to social inequality does not mean that the project determines the stakeholder. It is the responsibility of the project to be aware of the inequality underlying access to, and control over water and its manifestation in stakeholder participatory processes. Hence, more thought needs to be given to how interests are, or can be, identified, formulated by the stakeholders and understood by others. This is compounded by the lack of easy to use generic frameworks and tools to deal with equity issues in integrated water projects. And the few that do exist often cannot cope with the contextual complexity of social inequalities. Making the integration of social equity issues possible there where it is relevant, demands the topic to be placed centre stage. It is the development of such frameworks, instruments and tools that is partly the objective of the NeWater work package that deals with poverty and gender. Hopefully, the outcomes of NeWater may contribute to a better conceptual understanding of social inequality and more sensitive practice in water related development and research projects.

NeWater recognises the importance of social dimensions to achieve the project aim of identifying the means to support a transition to adaptive water management. Social equity is clearly identified and integrated in the project conceptualisation. This is definitely progressive compared to other water-related projects. In particular, the NeWater concept foresees gender sensitive analyses for the definition of problems and in the development and validation of potential solutions. This includes (Newater 2005):

- New technologies (gender appropriate),
- Systems approaches (gendered definition of water use and needs), and
- Management solutions (involvement of women in decision-making, gender empowerment for participation, gender sensitive communications and responses).

Hopefully such a gender analysis will avoid placing women and men in separate and distinct categories that always and everywhere have different needs or uses for water. Rather defining different uses and needs, it may be more fruitful to identify gender access to resources, and gendered access to decision making power, which may have both structural as well context specific dimensions.

Aquastress (www.aquastress.net)

The objective of Aquastress, which started in February 2005, is to develop stakeholder driven, European scale, comprehensive, multisectoral, integrated (institutional, socio-economic, technical) approaches for the diagnosis and mitigation of water stress. Aquastress wants to empower actors at different levels of involvement, at different stages of the planning process, and working at different spatial scales. The project is to result in prioritising actions that allow gradual improvement and flexibility to adapt to change. Aquastress will deliver guidelines to implement integrated water stress mitigation options at local, regional and European scale. It also promises to deliver a "culture change" in approaches to water stress (Aquastress 2004).

The project envisaged undertaking participatory activities and emphasis on participatory methods and tools. The Aquastress project developed a participation strategy to become fully "stakeholder driven" (Aquastress 2005). Similar to NeWater, the project learned that involving stakeholders in such a participatory process is far form easy (Aquastress 2006). For example, the general criteria it used for participatory processes is working with a representative quota. However, when applied, this may lead to the exclusion of weaker minorities in decision-making. Poorer sections of society making up a smaller relative percentage rarely have the means or the confidence to effectively participate to represent their issues based on their proportional population size. One possibility to compensate this lack would be to increase the number of representatives independently of the proportion to their actual population size. Here, the recognition of social inequality in IWRM would call for some sort of affirmative action.

Evaluating the recognition of social dimensions in the Aquastress project leads to the question to what extent the current approach to stakeholder driven processes integrates the understanding of social inequality when developing new approaches for diagnosis and mitigation of water stress. As already mentioned for the NeWater project it should not be forgotten that both the project members and stakeholders (being in most cases well educated and situated representatives of organisations and not ordinary citizens participating on a random basis) might be biased towards notions of social inequality.

Gewamed (www.gewamed.net)

Of the projects discussed so far, Gewamed is the only one exclusively addressing gender dimension of water management. Unfortunately, the project has just started and few results are available to be evaluated. The approach is based on generating and sharing knowledge through networking and cooperation by creating internet-based networks, and by organising workshops and conferences at the international as well as the regional level. It also hopes to develop a set of performance indicators of gender equity in the irrigation sector (Gewamed 2006). It will be interesting to see if such performance indicators allow for critical reflection on social inequality or if for the purpose of generic application they need to reduce complexities of inequality to simplifications based on generic notions of gender and poverty. If they do succeed, however, the next challenge lies in their application beyond the irrigation sector to include the WSS sector and at different spatial scales such as river basins.

One of the weak points of the GEWAMED project is the absence of a link with the local level and scrutiny of policy implementation. The reasons why for example gender analysis in local project and programme planning fails, often lies in disconnecting policy from the complex socio-economic interests and motivated behaviour and/or cultural traditions of the local context. Willingness to tackle gender inequality demands a fair amount of commitment by the local planning sector and the different stakeholders.

If GEWAMED succeeds in coming up with an open adaptable scheme or template for analysing gender inequality in IWRM policy and practice, future European water projects could significantly benefit from GEWAMEDs' results. This may be particularly interesting for both the NeWater (clear concern for gender) and Aquastress projects (same region). Unfortunately, these two projects may finish before GEWAMED is able to publish its results.

10.3.3 Summary of evaluation results and its relevance for IWRM

"*IWRM can establish the right policy framework and rules of the game – with equity as a key policy driver – as well as the necessary institutions to administer these rules* (GWP 2003). This statement argues that the right policy framework for implementing IWRM is required but that IWRM itself should influence such a policy framework on the long run.

One of the most important aspects coming out of the HarmoniCOP project is that it provides the means to manage a participation process by respecting diversity and therefore issues of social inequality. As already stated in Section 2 of this chapter, participation does not necessarily lead to gender sensitive water management, nor the recognising let alone alleviation of social inequality. The HarmoniCOP handbook addresses the problem of stereotyping and inadequate participation processes that – in the worst case - can be even more damaging than not having a participation process. Inequality is taken on as a methodological challenge. To tackle inequality itself and linking participation with meaningful action to overcome this inequality, the handbook could have included 1) an analysis of inequality so as to identify what the social and economic costs are for different participants, and 2) that participation is riddled with power differences and hence involves common and conflicting interests

The analysis of the social implications of the different water liberalisation scenarios conducted by the Euromarket project provides some insight into how liberalisation of the water sector may affect the poorer social strata of selected countries in Europe, but without analysing the underlying causes of poverty or how liberalisation could possibly alleviate these. The project suffers from a lack of gender-disaggregated data. It also suffers from an inadequate analysis of the complexities of community composition. A more comprehensive analysis of water pricing and the interrelatedness of social and environmental needs in relation to the cost recovery principle are missing. The conclusion of the project ends up being quite ambivalent.

The Intermediaries project is interesting with regard to the analysis of the role of intermediary organisation in stimulating changing social practices and introducing new technologies. It highlights the interconnectedness of different sectors and interests involved in water management that shape the implementation of IWRM policies, but does not investigate social inequality in communities, neighbourhoods and households. The involvement of the poor remained at the level of informants. The less formal or alternative organisations of these actors who have lost faith in the mainstream avenues for change were not considered. Participation remains a far more attractive exercise to those who see a considerable chance of their interests being attended to. If intermediaries are to have an impact in changing social practices by introducing new technologies it is essential to differentiate among all the actors involved or affected.

Social inequality needs to be more thoroughly investigated in the European context of water management, rather than defining the issue as one only relevant in the global south. Neither one of the case studies of HarmoniCOP nor any of the European case studies of NeWater specifically address poverty and equity. In the NeWater project the European researchers and stakeholders themselves could hardly imagine that a gender approach or the integration of poverty aspects may benefit their case study. Once a few women have higher positions in water management and function as stakeholders in the participation process, gender equity is felt to have been dealt with. This, however, is far too easy an excuse for not developing a more detailed analysis of different levels of social inequality in the practice and impact of contemporary water management. NeWater does recognise the importance of social dimensions to achieve the project aim of identifying the means to support a transition to adaptive water management. Social equity is clearly identified and integrated in the project conceptualisation. This does mean, however, that it is the responsibility of the project to be aware of the inequality underlying access to, and control over water and its manifestation in stakeholder participatory processes. Hence, more thought needs to be given to how interests are, or can be, identified, formulated by the stakeholders and understood by others. As to its gender component we urge the project to steer away from placing women and men in separate and distinct categories that always and everywhere have different needs or uses for water. More fruitful would be to identify gendered access to resources, and gendered access to decision making power, which may have both structural as well as context specific dimensions.

Similar to Newater, the Aquastress project claims that the strong participatory role given to stakeholders means that the inclusion of gender analysis in the research will not only depend on the individual interest of the researchers, but also those of the stakeholders too. Within the stakeholder-driven approach of the two projects the recognition of equity and gender during the analysis becomes at least partly dependent on the level of interest and biases of the respective stakeholders. To what extent the current approach to stakeholder driven processes integrates the understanding of social inequality when developing new approaches for diagnosis and mitigation of water stress remains unclear. Despite some shortcomings these projects are among the first to systematically integrate gender aspects, for example by using gender-disaggregated data and by recognising social differences through the means of extensive participation.

Gewamed has gender as its primary focus. By generating knowledge and developing gender indicators it may fill some of the mentioned gaps of other projects but little can be said at this point given the recent start of the project. Whether such indicators will resist easy simplification and incorporate the interlinkages between gender and other dimensions of inequality is unclear. GEWAMED itself is less concerned with the local level, which raises questions to what extent the contextual dynamics that reproduce gender inequality will be made explicit.

10.4 CONCLUSIONS AND RECOMMENDATIONS

The complexity and layerdness of IWRM demands recognition of similar complexity and layeredness of the society in which it takes place. Therefore the objective of this chapter was to provide an evaluation of EU funded research in light of contemporary insight in the social dimensions underlying water resources management. A consensus on a more integrated approach has opened doors to incorporate an analysis of the multiple social and economic dimensions of inequality that underlie the current challenges facing the water sector. We have provided such an analysis for three specific issues central to IWRM: water reallocation, water pricing and participation, and show that

(1) When taking social inequality in water management into consideration such reallocation demands new scrutiny in relation to food and income security;
(2) Water pricing will only work if it is flexible, context specific and highly sensitive to historic inequities and contemporary social dynamics.
(3) Participation is a highly complex and often problematic process and if it is to play a central role in IWRM, we need to understand and be willing to change the existing rights, roles and rules in so far that they allow social and material practices that reproduce inequality.

Furthermore, these issues highlight that IWRM can no longer neglect the social/material practices that underlie the various policies, institutions, and professions that make up the water sector. All these are shaped and influenced by gender and other forms of inequalities such as income, ethnicity, and religion. The review of the EU funded research projects has shown that incorporating social inequality is primarily a concern for projects funded in the countries of the global south. For the European context this is limited to countries in transition or in the Mediterranean region. Gender has been given more attention although only by a relatively small group. The projects discussed above show that more emphasis is placed on seeking approaches to overcome perceived inequalities rather than analysing their underlying causes. The projects that do explicitly attempt to address social inequality have found that participation is a far from self-evident process and demands reflexivity not only of the stakeholders but of the donors and facilitating agencies as well. In this respect increased migration may need more attention so that participating groups better reflect the make up of contemporary European society. This also calls for more detailed empirical research on the inter-linkages of income and water security, labour investment and water use, and develop gender and income disaggregated data to fully understand the implications and opportunities for IWRM.

An integrated approach to water management needs to take inequality seriously. This demands investment in both educational and research development that focus on inequality as a central concern of water management, not only in the global south but also in the global north. In the projects evaluated, social dimensions such as gender, poverty, and health are dealt with as points of concern rather than manifestations of inequality. Hence, little room is given to analysing the underlying causes or how these social dimensions are linked to structural impact on water management. Because IWRM is also very much a political process, its social dimensions need to be analysed as equally political.

Gender, income, and global inequality can thus no longer be treated as mere characteristics of a particular context, but rather as drivers for the political process that shapes IWRM.

This means that decision-makers have to assess their own circumstances and their own interests. This level of self-reflection is an important precondition for the requested social learning that is to facilitate the adaptiveness of management regimes (Section 1.4.3). It also means recognition of informal institutions and the need for creating awareness of, and respect for, different perceptions and beliefs of stakeholders in participatory processes. If systematically applied, social learning and the necessary analysis of the water management context (on the one hand the analysis of the governance structure, a detailed actor and institutional analysis and on the other the analysis of the natural environment and the influencing technologies) could also lead to a better recognition and understanding of social inequality.

The projects involving participatory approaches (all except Euromarket and GEWAMED) experienced that such a process is far from self-evident. They show that a concerted effort is needed to include all stakeholders; time and energy needs to be invested into developing a common appreciation of the objective —to "speak the same language" as it were— and have the same understanding of participation and the learning process. These cannot be assumed. This requires more than methodological trainings on participation, it requires reflexivity of all involved, including addressing power differences. This demands that social inequality, whether income, gender or otherwise is questioned and addressed from the beginning. New and more adaptive ways of resource management (NeWater 2006) need to be able to deal with different and often contradictory concepts of desired water management as well as understand how these concepts depend on contextual social and bio-physical particularities.

In addition to the evaluation of European projects with regard to social inequality, some attention should be given to the new gender action plan. One progressive aspect of the 6th European research framework program is that instruments, as Integrated Projects and Networks of Excellence, have to develop their internal Gender Action Plan as part of the work description. Besides annual status quo reports on gender, projects have to deliver at the end of their duration a final report explaining how they contributed to promote more gender equality. Being a positive innovation in research funding, inter alia with the objective to raise the number of women researchers in Europe, the effectiveness of this undertaking is questionable because no financial resources are provided to guarantee an implementation and follow-up. This concerns the monitoring and report writing each project partner is obliged to do, but also the implementation of measures that are defined in the plan. A better integration of the plan into the overall description of work allows that project partners have to bear the consequences if they disrespect the gender action plan as part of their contractual obligations. EU projects funded within the 6th framework programme, which included the obligation for each project to define its gender action plan, lack a coherent strategy to implement it. Although the final reports of the GAPs are not available (the FP5 projects did not have them, and the FP6 projects are far from being completed), it currently represents rather a toothless tiger adding to the bureaucratic burden of project partners. A thorough gender analysis, beyond a simple male-female dichotomy, that incorporates multiple dimensions of inequality may provide IWRM with the necessary interdisciplinary focus to affect a truly integrated approach to water management.

10.4.1 Recommendations for further research.

The following topics for further research are derived from the review of the above projects in articulation with our analysis of social inequality in the water sector. The topics below therefore address both the questions the projects have raised and the gaps they have left:

- A gender analysis is needed that investigates multiple dimensions of inequality of the social implication of IWRM policy reforms according to regional and hydrological variability in order to identify context specific and interrelated underlying causes of inequality.
- An analysis of global political drivers of IWRM policy incorporating an analysis of the relation of the global north with the global south is needed to identify to what extent global water policy directives are appropriate for different national or regional contexts.
- In addition to that, an analysis is needed of how the range of already implemented IWRM policy reforms (at different scalar levels such as national, regional, etc) impact on social inequality.
- Besides the need for disaggregated data on gender and income to fully understand the implications and opportunities for IWRM, data is needed on the inter-linkages of income and water security, labour investment and water use, legal pluralism of water and other linked rights or informal claims to land, or common property resources.
- Further, an analysis is needed of the composition of the epistemic communities and the decision-making bodies of the water sector itself. How accessible and socially reflective are they and what are the implications for water management, for instance in terms of possibilities for participation?

- Better understanding is needed of local and regional variations in decision-making procedures to identify constellations of power and mechanisms of representation to be able to improve participatory processes.

10.5 REFERENCES

Agarwal, B. (1994). *A Field of One's Own: Gender and Land Rights in South Asia.* Cambridge: Cambridge University Press.

Ahlers, R. (2000). Indirect Access and Informal Networks: Gender Issues in Irrigation, a Latin American Overview. In C. Tortujada (Ed.), *Women and Water Management, The Latin American Experience.* (pp. 203-217). New Delhi: Oxford University Press.

Ahlers, R. (2002). Moving in or Staying out: Gender Dimensions of Water Markets in a Mexican Irrigation District. In S. Whiteford and R. Melville (Eds.), *maamging a Sacred Gift: The Privatisation of Water Rights in Mexico.* San Diego: UCSD.

Ahlers, R. (2005). Gender Dimension of Neoliberal Water Policy in Mexico and Bolivia: Empowering or Disempowering? In V. Bennet, S. Davila-Poblete and M. Nieves Rico (Eds.), *Opposing Currents: The politics of Water and Gender in Latin America.* Pittsburg: University of Pittsburg Press.

Aquastress (2005). Deliverable 2.1-1 Common Framework for the Characterisation of Water Stress

Aquastress (2004). Description of Work (DoW).

Aquastress (2006). Design of AquaStress Questionnaire for Test Sites Survey of Stakeholders' Perception of Water Stress.

Arrighi, G. (2002). The African Crisis; World Systemic and Regional Aspects. *New Left Review, 15*(May /June), 5-36.

Arrighi, G., and Silver, B. J. (Eds.). (1999). *Chaos and Governance in the Modern World System*: University of Minnesota Press.

Bakker, K. (2003). *An Uncooperative Commodity: Privatizing water in England and Wales.* Oxford: Oxford University Press.

Baxter, P. T. W., and Richard, H. (Eds.). (1990). *Property, Poverty and People: Changing Rights in Property and Problems of Pastoral Development.* Manchester: Dept of Social Anthropology and International Development Center, University of Manchester.

Bond, P. (2004). Water Commodification and decommodification Narratives: Pricing and Policy Debates from Johannesburg to Kyoto to Cancun and Back. *Capitalism, Nature and Socialism, 15*(1), 9-25.

Cleaver, F. (1998). Choice, complexity, and change: gendered livelihoods and the management of water. *Agriculture and Human Values. Journal of Agriculture, Food, and Human Values Society. , 15*(4).

Cleaver, F. (1999). Paradoxes of Participation *Journal of International Development, 11*(4), 597-612.

Cleaver, F. (2000). Moral ecological Rationality, Institutions and the Management of Common Property Resources. *Development and Change, 31*, 361-383.

Cornwall, A. (2003). Whose voices, whose choices? Reflection on gender and participatory development. *World Development, 31*(8), 1325-1342.

Craps, M. (2003). Social Learning in River Basin Management, HarmoniCOP deliverable 3a, (www.harmonicop.info)

Deere, C. D., and Leon, M. (2001). *Empowering Women: Land and Property Rights in Latin America.* Pittsburg: University Of Pittsburg Press.

Escobar, A. (1995). *Encountering development : the making and unmaking of the third world.* Princeton, N.J.: Princeton University Press.

Euromarket (2003a). Analysis of the European Unions Explicit and Implicit Policies and Approaches in the Larger Water Sector. Deliverable 1.

Euromarket (2003b). Description of Work (DoW).

Euromarket (2005). Work Package 8: Analysis of the Social Implications of the Different Water Liberalisation Scenarios.

Frederiksen, H. D. (1996). "Water crisis in developing world: misconceptions about solutions". *Journal of Water Resources Planning and Management,, March/April.*

GenderandWaterAlliance. (2006). (http://www.genderandwater.org/page/2417).

Gewamed newsletter, Issue No. 1, June 2006, http://www.mio-ecsde.org/filemgmt_data/files/GEWAMED_Newsletter_1.pdf

Global Water Partnership (GWP) (2003). Poverty Reduction and IWRM. TEC Background Papers No.8, p.21

Gyawali, D. etal. (2006). *EU INCO water research form FP4 to FP6 (1994-2006); A critical review.* Luxembourg: Office for the official Publications of the European Communities.

Hoogvelt, A. (2001). *Globalization and the postcolonial world : the new political economy of development.* Baltimore, MD: Johns Hopkins University Press.

Jackson, C. (1998). Gender, Irrigation, and environment: Arguing for agency. *Agriculture and Human Values, 15*(4), 313-324.

Kohl, B., and Farthing, L. (2006). *Impasse in Bolivia.* London; New York: Zed Books.

Laurie, N. S. M. (1999). Globalisation, neoliberalism, and negotiated development in the Andes: water projects and regional identity in Cochabamba, Bolivia. *Environment and Planning A, 31*, 1401-1415.

Mamdani, M. (Ed.). (2000). *Beyond Rights Talk and culture talk: Comparative essays on the politics of right and culture.* New York: St. Martin's Press.

Mayoux, L. (1995). Beyond Naivety: Women, Gender Inequality and Participatory Development: Some Thorny Issues. *Development and Change, 26*, 235-258.

Mckinley, D. T. (2005). The struggle against water privatisation in South Africa. In B. Balanya, B. Brennan, O. Hoedeman, S. Kishimoto and P. Terhost (Eds.), *Reclaiming Public Water.* Amsterdam: TNI & CEO.

Mearns, R., Leach, M., and Scoones, I. (2000). *The institutional dynamics of community based natural resource management: an entitlements approach.* Sussex: Institute of development Studies.

Medd, W., Fay, C. and Marvin, S. (2003): Comparative Review of Water Sector Restructuring in the EU Accession States, Intermediaries Work Package 1, 2003;

Meinzen-Dick, R., and Zwarteveen, M. (1998). Gendered Participation in Water Management: Issues and Illustrations from Water Users in Asia. *Agriculture and Human Values, 15*, 337-345.

Meinzen-Dick, R. S., Brown, L., Feldstein, H., and Quisumbing, A. (1997). Gender, Property Rights, and Natural Resources. *World Development, 8*, 1303-1315.

Molle, F., and Berkoff, J. (2006). *Cities versus Agriculture: Revisiting Intersectoral Water Transfers, Potential Gains and Conflicts.* Colombo: International Water Management Institute.

Moss, T (2005*). New intermediary services and the transformation of urban water supply and waste water disposal systems in Europe.* Intermediaries detailed report, 2005.

NeWater (2006). NeWater Report Series No. 12, Deliverable 1.7.1, Framework for adaptive water management regimes and for the transition between regimes.

NeWater (2005a). Description of Work (DoW).

OECD (2002) Transition to full cost pricing of irrigation water for agriculture in OECD coutnries (No.COM/ENV/EPOC/AGR/(2001)62/FINAL).

Perry, C. (2001). Charging for Irrigation Water: The Issues and Options, with a case study from Iran. *IWMI Research Report Series, 52.*

Ridder, D., E. Mostert, H.A. Wolters (eds.) 2005. 'Learning together to manage together - Improving participation in water management'. *Handbook of the HarmoniCOP project.* Osnabrück: University of Osnabrück

Sullivan, C., Matin, N., Ridder, D. (2006). Rapid appraisal of poverty, gender and health: Application to the NeWater Case studies, (NeWater Deliverable D2.4.1).

Savenije, H.H.G. and P. van der Zaag (2002). Water as an economic good and demand management; paradigms with pitfalls. *Water International* 27(1): 98-104.

Spronk, S. (2006). *Roots of Resistance to Urban Water Privatization in Bolivia: The "New Working Class," the Crisis of Neoliberalism, and Public Services.* Paper presented at the Annual Meeting of teh Canadian Political Science Association., York University, Toronto, Canada.

Swyngedouw, E. (2005). Accumulation by Dispossession: Privatizing H2O. *Capitalism, Nature and Socialism, 16*(1), 81-98.

Toner, A., and Cleaver, F. (2006). *The evolution of community water governance: democratising access or localising inequality?* Bradford: University of Bradford.

Turner, K., Georgiou, S., Clark, R., Brouwer, R., and Burke, J. (2004). *Economic valuation of water resources in agriculture: From the sectoral to a functional perspective of natural resource management.* Rome: FAO.

UNDP. (2003). *Mainstreaming Gender in Water Management*. New York: UNDP.

Van der Zaag, P. (2005). 'Integrated water resources management: relevant concept of irrelevant buzzword? A capacity building and research agenda for southern Africa.' *Physics and Chemistry of the Earth* 30: 867-871.

Watts, M. (2004) Resource Curse?: Governmentality, Oil and Power in the Niger Delta, Nigeria, *Geopolitics* [Special issue] 9(1):50-80

Wester, P., and Warner, J. (2002). River basin management reconsidered. In A. R. Turton and R. Henwood (Eds.), Hyd*ropolitics in the Developing World: A South African Perspective.* Pretoria: African Water Issues Research Unit (AWIRU).

Young, R. (1996). *Measuring economic benefits for water investments and policies*. Washington D.C.: The World Bank.

Young, R. (2000). Thoughts on economic evaluation and the western policy wars. *Water Resources Update*(116), 85-90.

Zalik, A. (2004) The Niger Delta: 'Petro Violence' and 'Partnership Development'. *Review of African Political Economy.* Vol.31 No.101: 401-424.

Zwarteveen, M. (1997). Water: From Basic Need to Commodity. A Discussion on Gender and Water Rights in the Context of Irrigation. *World Development, 8*, 1335-1351.

Zwarteveen, M. (2006). *Wedlock or Deadlock? Feminists' attempts to engage irrigation engineers.* Wageningen: Wageningen University.

Zwarteveen, M., and Neupane, N. (1996). Free Riders or Victims: Women's Non-participation in Irrigation management in Nepal's Chhattis Mauja Irrigation Scheme. *International Irrigation Management Institute (IIMI)Research Report, 7.*

IWA Publishing's authorised EU representative for General Product Safety Regulations is Diane D'Arras, 15 rue Duret, 75116 Paris, France, e-mail: safety@iwap.co.uk.

Printed and bound by CPI Group (UK) Ltd, Croydon, CR0 4YY
12/05/2026
02108892-0020